Young People, Place and Identity

Young People, Place and Identity offers a series of rich insights into young people's everyday lives. What places do young people engage with on a daily basis? How do they use these places? How do their identities influence these ntexts? By working through common-sense understandings of young people's behaviours and the places they occupy, the author seeks to answer hese and other questions. In doing so, this book challenges and re-shapes nderstandings of young people's relationships with different places and lentities.

his book is one of the first to map out the scales, themes and sites engaged with by young people on a daily basis as they construct their multiple lentities. The scales, themes and sites explored here include the body, ighbourhood and community, mobilities and transitions and urban–rural tings and how these all shape and are shaped by young people's identities. ch chapter explores how social identities (such as race, gender, sexuality, ss, disability and religion) are constructed within particular contexts and influenced by multiple processes of inclusion and exclusion. These discussions are supported by details of the research methods and ethical issues involved in researching young people's lives. Drawing upon research from a range of contexts, including Europe, North America and Australasia, this book demonstrates the complex ways in which young people creatively shape, contest and resist their engagements with different places and identities. The range of issues, topics and case studies explored blends together original empirical research, theory and policy.

Individual chapters are supported by key themes, project ideas and suggested further reading. Details of key authors, journals and research centres and organisations are also included at the end of the book. This textbook will be pertinent for undergraduate and postgraduate students and academic researchers interested in better understanding the relationships between young people, places and identities.

Peter Hopkins is a Senior Lecturer in Social Geography in the School of Geography, Politics and Sociology at Newcastle University, UK. His research interests include: young people's geographies; religion and place; and urban geographies of race, ethnicity and religi

Young People, Place and Identity

Peter E. Hopkins

LONDON AND NEW YORK

First published 2010
by Routledge
2 Park Square, Milton Park, Abingdon, Oxon, OX14 4RN

Simultaneously published in the USA and Canada
by Routledge
270 Madison Avenue, New York, NY 10016

Routledge is an imprint of the Taylor & Francis Group, an informa business

Typeset in Times New Roman by
RefineCatch Limited, Bungay, Suffolk
Printed and bound in Great Britain by
CPI Antony Rowe, Chippenham, Wiltshire

British Library Cataloguing in Publication Data
A catalogue record for this book is available from the British Library

Library of Congress Cataloging in Publication Data
Hopkins, Peter (Peter E.).
Young people, place and identity / Peter E. Hopkins.
p. cm.
Includes bibliographical references.
1. Youth. 2. Identity (Psychology) in adolescence. 3. Place attachment–Social aspects.
4. Place (Philosophy)–Social aspects. I. Title.
HQ796.H659 2010
305.235–dc22
2009045572

ISBN: 978-0-415-45437-7 (hbk)
ISBN: 978-0-415-45439-1 (pbk)
ISBN: 978-0-203-85232-2 (ebk)

To all of my students, past and present

Contents

Illustrations

Figures

Images

Tables

Boxes

Acknowledgements

This book is developed from the 'Young People, Place and Identity' module that I teach at Newcastle University and the 'Childhood and Youth Geographies' course I taught when working at Lancaster University. I am very grateful to those students who have studied either module for their ideas, enthusiasm and support. In particular, thank you to Anna Boden, Heather Ponton-Brown, Michael Richardson, Ian Shaw and Alex Tan. Thanks to Anna and Ian for their useful comments on sections of an earlier draft, Alex for his help with images, and Michael for his research assistance. Raksha Pande has been an outstanding teaching assistant for the 'Young People, Place and Identity' module and I am very grateful for her enthusiasm and commitment. Photographs of some of the students who have studied this module are included between the chapters as a reminder that this book is for them.

I am extremely lucky to work in the School of Geography, Politics and Sociology at Newcastle University and a number of colleagues deserve a massive thank you for their encouragement and support. In particular, I would like to thank Anoop Nayak for his vast knowledge, understanding and critical insight into all aspects of young people's lives. Points of clarification and information from other colleagues were also very useful. Thanks here to Helen Jarvis, Alex Jeffrey, Nina Laurie, Nick Megoran, Alison Williams and Rachel Woodward. Conversations with Liz Trinder (now at Exeter University), Rob Hollands and Steph Lawler were also remarkably useful in helping me to clarify a number of topics and issues explored in this book.

Time spent as a visiting scholar at the National University of Singapore provided valuable thinking space and writing time to start working on this book. Thanks to Lily Kong for helping facilitate this visit and to Tracey Skelton for lending me some of her books whilst I was in Singapore. The critical comments of three referees really helped me to improve this collection and I thank the reviewers for their time, ideas and suggestions. Thanks also to Andrew Mould for commissioning this book and to both Andrew and Michael Jones for their work in seeing the book through to print. Many thanks also to my mother, Denise Hopkins, for her work in producing the index.

I also acknowledge the following permission to reprint materials. Box 1.2 from Danny Dorling, 'Anecdote is the singular of data' from D. Dorling, G. Smith, M. Noble, G. Wright, R. Burrows, J. Bradshaw, H. Joshi, C. Pattie, R. Mitchell, A.E. Green, A. McCulloch 2001, 'How much does place matter?' first published in *Environment and Planning* A 33(8) 1335–1369 and is reprinted with thanks to Pion Limited, London. Table 11.2 is from *The youth divide*: *Diverging paths to adulthood by* Gill Jones, published in 2002 by the Joseph Rowntree Foundation. It is reproduced by permission from the Joseph Rowntree Foundation. Table 12.2 is reprinted from Alister Scott, Alana Gilbert and Ayele Gelan (2007) *The urban–rural divide: myth or reality*. Aberdeen: Macaulay Institute with thanks to the Macaulay Land Use Research Institute. Table 12.2 is reprinted from 'Young rural lives: strategies beyond diversity', *Journal of Rural Studies* (2002) Volume 18 issue 2, pages 113–22 by Ruth Panelli, with permission from Elsevier. Image 2.1 is reproduced with permission of Caitlin Cahill.

Finally, special thanks are due to a number of other colleagues and friends. Rachel Pain, in particular, has been a constant source of encouragement, support and motivation, and has provided helpful feedback and thoughtful comments on some of the chapters. Sincere thanks are also due to Catherine Alexander, Nancy Bell, Caitlin Cahill, Malcolm Hill, Marilyn Keenan, Greg Noble, Elizabeth Olson, Katrina Roen and Nicola Ross.

1 Introduction

Whether it is due to the shifting nature of transitions to adulthood, the expansion of further and higher education or transformations in household composition and family life, young people are a regular topic of academic, popular and policy discussion and debate. They are a group of society whose experiences, behaviours and attitudes are usually misrepresented, often demonised and frequently distorted. Furthermore, it is clear that young people's experiences, actions and values vary across space and time, and according to the identities they display or are assumed to possess. It is at the intersections of these categories – age, place and identity – that this book focuses upon. Overall, the aim of this book is to explore the ways in which experiences of being young are mediated by different places and social identities.

Although I write this book as a geographer, it is important to recognise that 'geography is both an interstitial subject and an impulse to interdisciplinarity' (S. Smith 2005: 390). The location of geography simultaneously in the middle of, and in-between, a number of other social science disciplines results in this book being as much about social work, sociology and social policy as it is about gender studies, politics and culture. So, I see

> geography as an enterprise of relatedness whose vitality is secured by forging connections and crossing intellectual horizons; by pulling the world apart, reassembling it, and adding to it, in a variety of

> intriguing, ethically charged, sometimes surprising, and frequently controversial ways.
>
> (S. Smith 2005: 389)

The message here then is that this book is not only for geographers, and instead will appeal to those studying, researching or working with young people in a variety of different academic, educational and institutional settings.

Young people

In thinking about how age is constructed in society and across space, Rachel Pain (2001a) differentiates between chronological, physiological and social approaches to age. Chronological age is the number of years a person has lived, or the length of time that has passed since they were born. An approach informed by chronological age is applied to many aspects of young people's everyday lives. For example, a young person's chronological age frequently – if not always – determines the year group they are in at school, and influences whether or not a young person can legally consume alcohol, participate in an election by voting or standing for election, drive a motor vehicle, earn a salary or if they are allowed to get married or have a civil partnership. Many of these often rely on a young person being asked to prove their age through presenting a birth certificate, a passport or being asked their date of birth.

The second approach – physiological age – relates to issues connected with the physiological and bodily appearance of a young person and so is often based on factors such as perceived health, sense of medical well-being and general appearance. 'In contemporary Western societies the age of our physical body is used to define us and to give meaning to our identity and actions' (Valentine *et al.* 1998: 2). Third, social age refers to the social values, attitudes and beliefs that are held about people of particular age groups. For young people, this often involves them being assumed to display certain forms of behaviour, use specific places and possess particular values, beliefs and attitudes. Furthermore, it is important to note that these processes do not only influence young people and instead shape the experiences of people of all ages. Like young people, older people are expected to possess particular values, use specific places and display certain behaviours and attitudes.

The experience of being a young person is therefore likely to vary according to the particular approach towards age that is being implemented in different places and times and by various institutions, groups or individuals. As a result of this – and like people of all ages – young people often experience ageism. Bill Bytheway (1995) sees *ageism* as the result of the ways in which assumptions are made about people having other things in common alongside their chronological age. However, it is important to note that ageism is not necessarily equivalent to racism or sexism in the sense that young people have been children and will get older, meaning that ageism is a constantly shifting phenomenon that changes through time.

The intersection of these different approaches to age in young people's lives often has powerful influences over how they construct and contest their identities in different places and times. The diversity of what it means to be a young person is highlighted here by Gill Valentine (2003: 38):

> Indeed, the terms 'youth' or 'young people' are popularly used to describe those aged between 16 and 25, a time frame that bears no relation to diverse legal classifications of adulthood. While children aged 5–16 should be at school, young people aged 16–25 may be at school, college or university, other forms of vocational training, in paid work, unemployed, doing voluntary work, travelling and so on. They experience far fewer spatial restrictions than their younger peers because it is easier for a young person in their late teens or early twenties to pass as 'older' than they actually are in order to gain access to places such as clubs and bars from which they might otherwise be excluded. Young people also usually have access to some form of income independent from their parents, some live away from the parental home on a temporary or permanent basis, and their independent mobility can be enhanced by the possession of driving or motorbike licences. As a result their lives are less obviously circumscribed by parents, teachers or other adults.

Mary Jane Kehily (2007a: 3) notes that 'definitions of "youth" in Western societies usually refer to the life stage between childhood and adulthood, the transitional period between being dependent and becoming independent'. So, young people are often associated with the chronological age range of 16–25, located uncomfortably between – yet simultaneously overlapping – childhood and adulthood. As noted above, young people are less spatially restricted than younger children yet may not possess a number of the key qualities associated with adulthood. This picture is further complicated by the diversity of everyday situations young people may experience. Unlike children they may not be at school, although some still are and more still are involved in learning in some form or another. They may be very independent in terms of their housing and financial circumstances, although many young people still rely on family support for some – if not all – of their living expenses. This ambiguous phase is further complicated by the fact that being a child is ratified in the United Nations Convention on the Rights of the Child as being a person of up to 18 years of age, and adulthood – and the vast majority of the rights and entitlements associated with it – are available when a person turns 18 (although certain entitlements also vary further depending on the spatial context in which young people find themselves).

For Johanna Wyn and Rob White, young people are not adults and are seen as being in need of guidance. For them, youth is a relational concept, highlighting the power relations that are part of the experience of being young:

> Youth is a relational concept because it exists and has meaning largely in relation to the concept of adulthood. The concept of youth, as idealised and institutionalised (for example in education systems and welfare organisations in industrialised countries) supposes eventual arrival at the status of adulthood. If youth is a state of 'becoming', adulthood is the 'arrival'.
>
> (Wyn and White 1997: 11)

As highlighted in Box 1.1, youth is seen in relation to adulthood, highlighting the relationality that is a key aspect of being a young person.

Although some young people are also children and many young people are defined by the fact that they are not adults, there are young people who are adults and embody many of the qualities characteristic of adulthood. This diversity is probably one of the main reasons why studies of youth are characterised by a diverse range of theoretical approaches.

The young people discussed in this book are generally aged 16–25, although some may be younger or older depending on the particular study, situation or issue being discussed. In thinking about young people, place and identity, you may regard yourself as a young person and so will have an ideal vantage point to think critically about what being a young person means. However, you may be older and may feel

Box 1.1

Notions of youth and adult

Youth	Adult
Not adult/adolescent	Adult/grown up
Becoming	Arrived
Presocial self that is emerging	Identity is fixed and given
Powerless and vulnerable	Powerful and strong
Less responsible/irresponsible	Responsible
Dependent	Independent
Ignorant	Knowledgeable
Risky behaviours	Considered behaviour
Rebellious	Conformist
Reliant	Autonomous

Adapted from Wyn and White (1997: 12)

that you are no longer a young person. An important issue to consider here is highlighted by Chris Philo (2003: 9) when he notes:

> 'all adults have at an earlier time of their lives *been children*. We have all 'been there' in one way or another, creating the potential for some small measure of empathy – some sense of recognition, sharing and mutual understanding, even if slight – with the children whom we encounter in our adult lives.

Likewise, even if we do not consider ourselves as being a young person, we have all been young people and so we have all experienced youth.

Identities

Just as 'young people' or 'youth' are complex terms with many meanings and definitions, the same too can be said about identity. As Steph Lawler (2008: 2) notes 'it is not possible to provide a single, overarching definition of what it is, how it is developed and how it works. There are various ways of theorising the concept, each of which develops different kinds of definition.' It is not therefore possible to answer the question, 'What is identity?' However, there are certain key facets to identity that can be acknowledged. Starting with the Latin root of the word identity – *identitas*, from *idem* – we learn that identity has a connection with being 'the same' as something else. This implies both absolute sameness as well as distinctiveness (Jenkins 2004). So, 'approaching the idea of sameness from two different angles, the notion of identity simultaneously establishes two possible relations of comparison between persons or things: *similarity*, on the one hand, and *difference*, on the other' (Jenkins 2004: 3–4). Similarity suggests that 'we share common identities' (Lawler 2008: 2) – say, if we are young, white and male – yet we are also different as well. Yet, being similar and being different from another person are open to negotiation, management, disagreement and contestation. As key components of identity, similarity and difference are dynamic and can change over time, taking on new meaning or forms, or being reinterpreted and renegotiated. Likewise, other senses of *similarity* or *difference* may not change much at all over time and may be more durable, more consistent and less flexible depending upon how they are articulated, where and when.

Identity is therefore about

> the ways in which individuals and collectivities are distinguished in their social relations with other individuals and collectivities. It is the systematic establishment and signification, between individuals, between collectivities, and between individuals and collectivities, of relationships of similarity and difference.
>
> (Jenkins 2004: 4)

Many would therefore classify young people or youth as an identity in that they recognise ways in which young people are distinguished in their social relations from other groups in society. Moreover, identity is also about 'our understanding of who we are and of who other people are and, reciprocally, other people's understanding of themselves and of others (which includes us)' (Jenkins 2004: 5). So, a young person may be similar in particular ways to another young person, yet different in others, and overall they may be very different and have only their age in common. Identity is therefore *relational*, and relies on establishing senses of difference and similarity between different individuals and across and within different social groups. Part of this relationality often means that 'varying and often contradictory identities must be managed' (Lawler 2008: 3). Furthermore, all identities are mutually exclusive and rely on not being another identity, drawing upon social binaries such as man/woman, black/white and heterosexual/homosexual. Identities rely on disidentification as well as identification, and so are oppositional, sometimes resulting in tension between identities and exclusion of what they are not.

Identity does not just exist per se as it is given meaning through being taken up by people and articulated in particular ways. The verb *identify* therefore gives agency to the noun *identity*. Identity is therefore also about the processes of identification that people engage with in order to identify in particular ways or dis-identify in others. So, it is about the simultaneous creation, maintenance and protection of particular categories, and the rejection, renunciation and disavowal of other categories. These categories – or identities – often relate to significant divisions within society and some of the most common relate to age, gender, sexuality, class, dis/ability and race/ethnicity (see Table 1.1). These different identities are 'socially produced, socially embedded and working out in people's everyday lives' (Lawler 2008: 8), and, as Kevin Dunn (2001: 292) observes, 'social constructions of identity are given life through their articulation. Through repetition they can achieve a remarkable durability. In a sedimentary-like process the reinscription of social constructions . . . can come to be widely accepted as unproblematic, and as a natural given'. Likewise, Susan Smith (1999: 12) notes:

> processes of social categorisation do not occur 'naturally'. They are a product of how power and resources – which may be real (money, cars, homes) or symbolic (a question of how people think, and what they take for granted) – are struggled over and manipulated.

Identities are therefore constructed through social relations, articulated in particular ways and replicated by individuals and groups. The tenacity, persistence and governance of specific identities often lead to them being interpreted as genetic, biological or natural, yet identities are constructed through social relations and people's everyday behaviours. It is important to recognise that – as Jan Penrose (1995: 401) observes – 'individual identity formation' is relevant to all human beings, and whilst categories are important, we also need to look at the processes that create, sustain and support them. So, in thinking about identity and the ways it is given meaning through social relations, we also need to think about the *processes* that create particular identities and make them matter in different ways (see Table 1.1).

When focusing upon identity, there is a tendency to think of singular identities such as gender, race or class existing on their own. In many ways this makes sense – take, for example, youth. There are complex social processes at play that work to influence the ways in which youth identities are constructed and much work across the social sciences for a number of decades has made important interventions in this field

Table 1.1 ***Identities, relationalities and processes***

Identities	*Relationalities*	*Processes*
To identify or be identified	Similar to and/or different from	Social and cultural mechanisms and discriminations for sustaining particular identities
Age – connected with a person's chronological, physiological or social age (e.g. child, teenager, youth, adult, older person, pensioner)	Being a child and therefore not an adult, and vice versa	*Ageism* works to create and sustain assumptions about aged individuals and their behaviours, attitudes and values
Class – relating to the material and economic positioning of individuals (e.g. working-class, middle-class, upper-class)	Being middle-class and therefore not working-class, and vice versa	*Classism* operates to stigmatise particular social classes and associate them with disgust or distaste
(Dis)ability – identities connected with a person's perceived mental and physical capacity and mobility (e.g. visually impaired, wheelchair user, able-bodied)	Being able-bodied and therefore not disabled, or being a wheelchair user and not being regarded as able bodied, and vice versa	*Ableism* works to stigmatise disabled people as medically inferior, and *disableism* stereotypes disabled people based on assumptions relating to their capabilities
Ethnicity – identities relating to people's heritage and cultural backgrounds, encompassing factors (e.g. diet, belief, dress and language)	Being accustomed to particular cultural or ethnic practices (such as eating specific foods in a particular way, or adopting specific styles of fashion), or not doing this, or vice versa	*Ethnicisation* works to reinforce negative stereotypes about particular ethnic practices (often interrelated with racism and *racialisation*
Gender – identities associated with the socially constructed characteristics connected with being male or female (e.g. masculinities, femininities)	Being male and therefore not being female, or vice versa	*Patriarchy* upholds a system where men dominate women, and *sexism* works to maintain these identities and their associated inequalities
Race – identities relating to the socially constructed qualities associated with people based on skin colour, facial structure or type of hair (e.g. black, white, mixed-race)	Being black and therefore not white, or vice versa	*Racism* operates to maintain a hierarchy in which non-white people are excluded and vilified. *Racialisation* works to associate particular behaviours, places or traditions as related to race
Religion – identities connected with a person's religious beliefs or affiliations (e.g. Muslim, Sikh, Christian, pagan or agnostic)	Being Christian and therefore not Sikh, or vice versa	*Religious discrimination* operates to associate particular forms of belief or non-belief with aversion and distrust
Sexuality – identities associated with an individual's sexual orientation (e.g. bisexual, gay, heterosexual, lesbian, gay, queer, transgender)	Being heterosexual and therefore not lesbian, or vice versa	*Heterosexism* supports processes where heterosexuality is seen as normal, and anything else abnormal. *Homophobia* works to sustain a hierarchy in which non-heterosexuals are stigmatised and marginalised

Although these are the main identities commonly referred to in the social sciences, arguably, other identities include those associated with:

Indigeneity – identities associated with a person's status as belonging to particular indigenous groups or cultures (e.g. Maori, Aboriginal, Sami)	Being indigenous and therefore not part of a colonising group, or vice versa	*Colonisation* works to subvert, exclude and marginalise the experiences, practices and values of indigenous groups, normalising those of the colonisers
Locality – identities connected with living in a particular street, neighbourhood, community, city or part of a country	Being from a particular neighbourhood and therefore not from another community	*Territorialism* works to reinforce local boundaries and differences through resource allocation, stereotypes or rumour to create tension between different areas or districts
Size – identities connected with particular body sizes such as small, tall, skinny, fat, overweight, large or 'normal'	Being fat and therefore not being thin, or vice versa	*Sizism* operates to sustain a system whereby people who are not thin are regarded as lazy, arrogant and uncooperative

(France 2007, Furlong and Cartmel 2007, Hollands 1990, Skelton and Valentine 1998, Willis 1977). Many everyday assumptions and stereotypes about young people lead to assumptions being made about them being rebellious, out of control or a threat to the moral order of civilised society. These stereotypes are reinforced through the agents of socialisation – such as the family, school, the media and the workplace – that have exceptional influences over how people – of all ages – live their lives. These processes are also shaped by the ways in which young people are constructed in *relation* to both children and adults, meaning that young people are often stereotyped as not being as independent, mature or as sensible as adults, whilst not also being as cute, innocent and as vulnerable as younger children. Yet, young people may embody these various different qualities at different times as they may be innocent, vulnerable and dependent upon their family for support, or they may be very independent, working, bringing up a family or caring for parents who are unwell.

However, focusing on young people as a single category of identity tell us only one part of a much larger story. As Steph Lawler (2008: 3) observes:

> No one has only one identity, in the sense that everyone must, consciously or not, identify with more than one group, one identity. This is about more than combining multiple identities in an 'additive' way . . . race, gender and the rest interact (as do all forms of identity), so that to be a white woman is not the same – in terms of meaning or experience – as to be a black woman. Different forms of

identity, then should be seen as interactive and mutually constitutive, rather than 'additive'. They should also be seen as dynamic.

So, young people also – consciously or not – connect with and take up social identities other than age and so are also racialised, classed, gendered, sexualised and so on in particular ways in different spaces and times. These various identities do not simply add on to each other in a numerical sense but instead interact, overlap and play off against each other in complex ways – they intersect – and seeking to understand what happens at these different intersections is a key challenge for social scientists and a task that I hope this book might contribute to. A central issue in studying young people, place and identity is to consider the ways that social identities and difference are recognised and critiqued, and the power relations that are a part of this.

So identities are about processes, relationalities and intersections. Focusing upon the processes associated with identity means that we can look at how (1) young people actively communicate their identity, (2) how it is interpreted by others, (3) how it is variously negotiated or (4) how it is represented or performed (Panelli 2004). Identities also shape and are shaped by particular geographical contexts and settings (5) (see Figure 1.1). The intersections and interactions between these different processes and relationalities in the context of young people, place and identity are a key focus of this book, and are more or less relevant in different situations and so are raised – where appropriate – in different chapters.

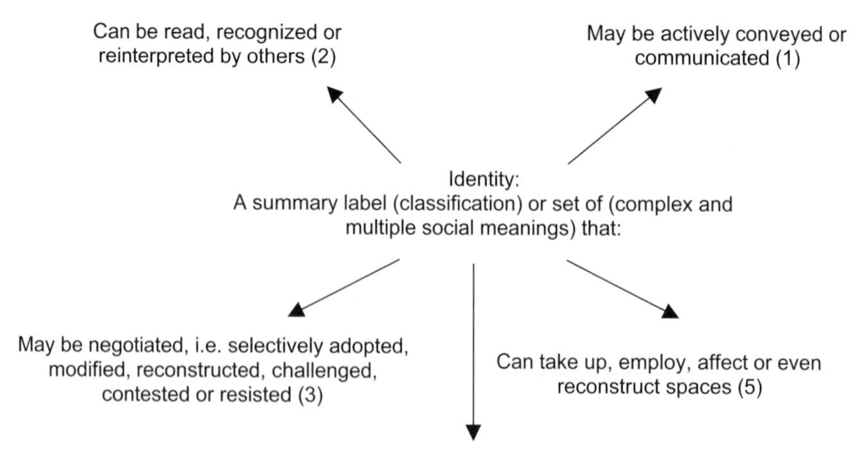

Figure 1.1 Processes of identification.

Place

Place tends to be understood as a unique location that is connected to other places but is also self-contained and distinctive. As a result of the significance of a place to the people residing within it, a 'sense of place' can develop (Massey 1994). Place has been increasingly understood as subjective, relational and as something that can be made and promoted. Rather than having fixed boundaries, place is now recognised as having open and permeable boundaries, shaped by complex webs of local, national and global influences and different social and cultural flows and processes. A key argument of this book is that the places used, inhabited and associated with young people matter; geography matters.

Consider Danny Dorling's reflections about his journey to school from age 6 through to 18 (Box 1.2) and the different ways in which he began to learn at age 12 that place really mattered. It is clear that the route into and out of the pedestrian underpass said something about where you lived, what kind of person you were, and where you might be heading in the future. The young people's use of the underpass with the adults using the main street also speaks to other forms of social division and inequality. Who you met in the subway also mattered and identities such as age, gender, race and size are all mentioned as factors that could determine whether or not a young person was bullied, beaten up, ignored or avoided. Also, in the article that this extract is taken from, Danny Dorling mentions in a footnote that 'children going to private school did not use the subway – I think they must have travelled by car', pointing to particular class divisions in young people's use of different places and spaces. Clearly, the particular ways in which young people's identities are conveyed, read, negotiated and performed – in particular places – really matters.

Just as young people's identities will influence and be influenced by particular places, the specific places or locations that young people find themselves in also act as an important marker of identity and sense of identification. People are often identified according to the particular nation they affiliate with, the neighbourhood or street they live in or the school they attended. Like other identities, those associated with place and localities can be better understood by appreciating the processes, relationalities and intersections associated with them. Similarity may be established as a result of occupying similar places such as attending the same school or youth group. Yet differences, such

Box 1.2

The importance of place . . .

From the ages of six to eighteen I got to school on foot or bike through a subway [pedestrian underpass]. The subway . . . was entered by steep sloping ramps, was damp, long, and usually very dark as the lighting was smashed. The walls were graffiti covered. Among the names of children I knew, or thought I knew, were nastier slogans. The National Front were strong at the time and appended a swastika to their two joined-up initials. Swastikas are easy to spray paint. Most importantly, however, to me, the subway was curved. You couldn't see who was inside when you were going down into it. Once inside you could not see who was round the corner. Only when coming up again to the surface could you see light. And it mattered who you bumped into, depending on where you were entering and exiting from, how old you were, how small you were, whether you were a boy or a girl, black or white.

The roads divided a large council estate to the north, from the 1930s semidetached housing of my estate to the east, from the picturesque 'urban village' to the south, and the mixed development of the west. The subway connected these four corners and was where children in the 1970s and 1980s met between these different worlds. Adults often preferred to risk crossing the dual carriageways [divided highways]. Each morning and evening what appeared to be thousands of men cycled four abreast from the council estate round the roundabout to work in a large car factory half a mile south. Mums with prams would walk their smallest children to schools in each direction from each direction, over the surface, crossing the roads. Older children and teenagers (and a few pensioners – too slow for the roads) would go underground. There were half a dozen primary/junior schools and two middle schools. Which you went to said something about where your parents thought you were coming from and going to. Everyone I met in the subway ended up at the same secondary school at age thirteen. But by then where they were going to next (how they would add to the sediment of society) was often largely decided.

Why did I begin to learn that place mattered at age twelve? Because it was then that I began to notice who came into the subway from where and by which exit they left (in effect, where they lived and where they went to school). What happened to my neighbours six years later appeared, to me, to depend acutely on children's comings and goings in the subway earlier . . . If you got a kicking at age twelve because you came in from the wrong entrance of a subway, you learnt quickly that place matters.

Adapted from Dorling (2001: 1336–8)

as being a member of an opposing football team or living in a different neighbourhood, may be prevalent. This connects with the ways in which places are relational to other places. So, specific locations are constructed in ways that establish connections or disconnections with other places, and these differences and similarities often have to be managed in ways which may be contradictory. Social processes, such as those associated with the media or neighbourhood rumour, work to stereotype particular places – and the people living in such communities – in negative or positive ways. To identify as belonging to particular places, similar to identifying with specific social groups, or disidentifying with certain locations and disavowing other identities are processes, relationalities and intersections associated with identity. Clearly then, just as age, gender and race are identities, so too are locality and place.

Thinking about youth and place as socially constructed categories and forms of identification, there are three main ways we can appreciate the relationship between these categories (see Figure 1.1). First – and as Rachel Pain (2001a: 151) has suggested – 'people have different access to and experiences of space and place on the basis of their age'. This means that young people's experiences of place will vary according to their age, and so they may be granted access to particular places and refused entry to others. As such, spatial patterns are a reflection of social relations and vice versa (S. Smith 1999). Second, 'spaces have their own aged identities, which have implications for those who use them'. Places such as the university campus or the nightclub are locations that have strong associations with youth and so this will influence who chooses to use such places and their senses of belonging – or not – in them. Here, the organisation of space is mediated by social and political constructions (S. Smith 1999). Third, 'people may actively create and resist particular age identities through their use of space and place', and so young people may seek to actively challenge or resist their exclusion from or dis-association with particular places through using place and articulating their age in a particular manner. This approach allows for a re-imagining and a redefining of what the world is like, meaning that young people can reclaim marginalised spaces and make them into something new (S. Smith 1999).

Related to this, Elizabeth Kenworthy Teather (1999: 2) has suggested that we can look at place as:

1 the way 'we move from one location to another in the course of a day' – 'the space of place';

2 'activity space' – 'activity spaces comprise our many communities of interest, such as political lobby groups or recreational societies';
3 'each of us is positioned or located relative to others. Our positionality depends on who we are – our identity';
4 'discursive space' – 'a set of mental attitudes and conventions held by members of the public'.

Different places – or sites of social relations – are used to structure this book in order to emphasise the importance of place to young people's lives and other identities. Although these different scales, sites and themes are taken individually, they clearly interrelate and interconnect with each other, shaping, moulding and overpowering each in different ways and at different times. But, by focusing on each individually and gaining a detailed understanding of the ways in which each relates to youth and identity we hope to help build a sophisticated understanding of young people, place and identity.

Frameworks for studying young people, place and identity

Three concepts can be identified to help appreciate the relational geographies experienced at the meeting point of young people, place and identity: lifecourse, intergenerationality and intersectionality. Indeed, it has recently been proposed that such concepts could be usefully applied to understanding the experiences of people of all ages and not only young people themselves (Hopkins and Pain 2007, see also a response from Horton and Kraftl 2008).

Lifecourse

The lifecourse approach is one of the most common frameworks used when studying young people and 'involves recognition that, rather than following fixed and predictable life stages, we live dynamic and varied lifecourses which have, themselves, different situated meanings' (Hopkins and Pain 2007: 290). Or, in a metaphorical sense, Jenny Hockey and Alison James (2003: 5) note that lifecourse has 'been adopted as a way of envisaging the passage of a lifetime less as the mechanical turning of a wheel and more as the unpredictable flow of a river'. So, rather than focus on chronological aspects of age such as reaching adulthood when turning 18, applying a lifecourse approach to young people's lives recognises young-personhood as a dynamic and contested phase of life's journey: 'different stages of the lifecourse are socially constructed' and 'these constructions have significant implications for the use of space' (Pain 2001a: 141). This phase may have distinct qualities alongside features that overlap with over stages, phases or passages over an individual's lifecourse. As I have noted elsewhere:

> A young person's life course trajectory towards adulthood often involves negotiating a range of transitions: school to college/ university to work; parental home to shared accommodation with peers to their own home; child of a family to partnership/ cohabitation with partner with children; 'pocket money' income to part-time work/temporary income to full salary; and general economic dependence through semi-dependence to full independence (Jones 2002: 2; see also Hill and Tisdall 1997).
>
> (Hopkins 2006: 240–1)

The popularity of a lifecourse approach to understanding young people, place and identity is emphasised in Chapter 11 where I explore youth transitions as a key aspect of young people's journeys through life.

Intergenerationality

Focusing upon relations, interactions and tensions between and within different generational groups, the concept of intergenerationality can also help us to appreciate the experiences of young people. Here, the focus is upon the sense in which young people are part of a particular generation and therefore are in a different generation from other people, such as their parents or grandparents. Intergenerationality therefore relies on the concepts of similarity and difference mentioned earlier, but also recognises that young people are more than just young people due to their relations and interactions with different generations. Robert Vanderbeck (2007) has proposed that 'intergenerational geographies' could be useful in understanding experiences of age as such an approach helps to highlight the role of space in the facilitation or restriction of contact between the generations and how this connects with issues of contact, conflict or cohesion. The use of intergenerationality as an explanatory device has been questioned (Horton and Kraftl 2008); however, it has been noted that although 'there are times and places when intergenerationality might not figure at all in explaining particular aspects of people's lives', there are 'times and places when it can figure very powerfully'

(Hopkins and Pain 2008: 290). The significance and importance or absence and silence of intergenerational relations are considered at various points throughout this book, particularly in relation to thinking about young people's negotiations of home in Chapter 5.

Intersectionality

A third concept that helps to understand young people, place and identity is intersectionality. One of the first people to use this term was Kimberele Crenshaw who employed intersectionality in her analysis of the various ways in which gender and race influence the complex nature of black women's experiences of employment. She notes that she used intersectionality

> to illustrate that many of the experiences Black women face are not subsumed within the traditional boundaries of race or gender discrimination as these boundaries are currently understood, and that the intersection of racism and sexism factors into Black women's lives in ways that cannot be captured wholly by looking at the race or gender dimensions of those experiences separately.
>
> (Crenshaw 1993: 1244)

Intersectionality is useful as it 'foregrounds a richer and more complex ontology than approaches that attempt to reduce people to one category at a time' and 'indicates that fruitful knowledge production must treat social positions as relational' and so 'is thus useful as a handy catch-all phrase that aims to make visible the multiple positioning that constitutes everyday life and the power relations that are central to it' (Phoenix and Pattynama 2006: 187). So, as highlighted above, identities are not simply totalled up to reach a fixed outcome and instead intersect and interact in diverse ways, sometimes complementing each other, sometimes overpowering one another or sometimes coming into conflict. Intersectionality is therefore useful for helping us to consider who else young people are. Furthermore, the concept contributes to considering the ways in which young people's identities 'change, transform, or remain the same as they intersect and interact with other markers of social and cultural difference, in different places and times' (Pain and Hopkins 2009: 88).

Placing young people

Although much work about children and young people claims that concepts such as child or young person did not exist prior to the seventeenth century (Aries 1960), this has been challenged by academics on methodological grounds. 'Youth as a linguistic category did not exist, but boundaries between different age groups in pre-modern society did' (France 2007: 6). Alan France explains that, by taking a historical perspective on the topic of youth, it is clear that age boundaries between groups did exist, although these were often blurred and fluid according to local experiences of family, community and work. Furthermore, there were specific festivals for young people, and rites of passage between children and adulthood also existed. Also, during this time, there were anxieties about young people's place in society including concerns about their collective behaviours, concerns about the time they spent living away from home before being married and concerns about those who failed to take up work (France 2007).

The eighteenth century brought with it industrialism and transformations in social, economic and political life associated with it. This was accompanied by migration from rural to urban areas and changes in employment opportunities and patterns. Arguably, industrialism also restructured and transformed family life resulting in a bolstering of the significance of family. As Alan France clarifies, the influences of this phase 'on the meaning of youth and the youth question was significant. What we see is the early establishment of the making of the modern phase called youth' (France 2007: 9). These transformations worked to heighten concerns about how to regulate and control young people, with specific anxieties about youth crime increasing. Such concerns resulted in the state taking on the challenge of controlling young people gradually leading to additional legislation associated with everyday lives and behaviours of young people.

> The state then started . . . to universalise the youth phase as transitional and age-bounded. Youth became separated from both childhood and adulthood and existed in the new infrastructure constructed by the state. By the middle of the nineteenth century, youth, as a part of the life course, was well and truly established.
>
> (France 2007: 12)

It is evident that these changes in society and the economy – and those which have followed it – are interconnected with the ways in which academic researchers have explored young people's lives. So, although young people have been the focus of academic research for the vast majority of the last century, the volume and diversity of this work mirrors broader changes in society and perceptions about the place and behaviours of young people. Mary Jane Kehily (2007b: 12) observes:

> Large scale socio-economic changes in Europe and North America such as industrialisation, mass education and legislation regulating child labour created the conditions for children and young people to be separated, to some extent, from adults. Schooling, for example, organises children and young people into age-based cohorts, subjects them to similar experiences and, of course, delays the onset of economic activity.

These multiple changes resulted in specific responses from social scientists to understanding young people's lives. For example, the sharp increase in urbanisation led sociologists in the Chicago School to explore the behaviours of deviant youth in the city (France 2007, Heath *et al.* 2009). Furthermore, the reinforcement of compulsory schooling offered educational researchers the opportunity to explore the complexities of young people's experiences of different educational environments, and the collapse of the youth labour market led to serious concerns about young people's transitions from school to work, leading to research about youth transitions. These multiple responses – by society and by the academy – results in youth studies being a field characterised by much diversity. Sue Heath *et al.* (2009: 9) note that 'youth studies is a broad church; it embraces research on all aspects of young people's lives, and youth researchers are to be found across all social science disciplines'. The diversity of disciplines and perspectives that researchers bring to their work about young people means that there is now a range of well-developed traditions within youth research (see Box 1.3).

This book draws upon research with young people that sits across all of these traditions; however, given the focus upon place and identities, it draws more upon particular traditions compared with others. Drawing mostly upon research within the 'social and cultural geography' tradition, it also takes into account research based within other traditions such as cultural studies, youth transitions, educational research and feminist youth research. Arguably, many studies also draw upon and contribute to more than one tradition. Overall then, in

Box 1.3

Traditions within youth research

Developmental psychology

There is a long tradition of scholarship about young people within development psychology which tends to focus on adolescence as a phase in people's social and psychological development. Work here tends to use survey techniques and experimental methods, with some recent work moving towards incorporating qualitative methodologies. Overall, work within this tradition tends to see adolescence as a troubling and problematic phase of the lifecycle.

Educational research

Researchers working within educational research have contributed much to understanding young people's lives, particularly in terms of how young people experience different learning contexts, including both institutional and non-institutional settings. Work within this field draws upon a broad range of methodological approaches but Heath *et al.* (2009: 10) note that, in particular, educational research can 'be justifiably proud of its rich tradition of ethnographic studies of school life'.

Cultural studies

Cultural Studies has it origins within studies about young people, with work in the USA coming from the Chicago School and in the UK from the Centre for Contemporary Cultural Studies at the University of Birmingham. Research about young people within cultural studies often draws attention to popular culture, such as pop music, media and leisure. Research here tends to rely on the use of ethnography and other qualitative methods as mechanisms for accessing the lived experiences of youth culture.

Youth transitions research

Alongside research about youth (sub)cultures, research about youth transitions forms one of the strongest bodies of scholarship about young people's lives. This work tends to focus on key transitions, such as those from school to work, and is often linked with policy changes and debates. It emerged as a result of the collapse of the youth labour market in the 1970s and 1980s. Some work within this field draws heavily upon quantitative methods through the use of large scale surveys and longitudinal analysis of various data sets, whilst in recent years there has been an increase in qualitative work that uses narratives and biographies to access young people's experiences of transition.

Social and cultural geography

More recently, and in particular over the last ten or 15 years, a new theme of youth research has emerged that draws attention to the spatialities and temporalities of young people's lives. This work tends to rely on qualitative methods.

(*Continued overleaf*)

Feminist youth research and 'girl studies'

As much of the earlier work about young people was specifically about young men, feminist scholars have worked to include the voices and experiences of young women in youth research. There is now a strong body of work within youth studies which takes a feminist approach, often drawing upon qualitative research and seeking to empower young women.

Adapted from Heath *et al.* (2009: 10–13)

placing young people, it is important to think critically about the historical development of what it means to be a young person, the social and structural conditions which have shaped how young people are perceived and the ways in which academic research has responded to these issues by drawing upon and informing particular traditions of youth research.

How to use this book

Although I hope the book presents an overall narrative about young people, place and identity, I have not written it with the intention of it being read from cover to cover and instead expect that certain chapters or sections may be more relevant than others depending on your interests and motivations. The main points of each chapter are summarised in a 'key themes' section of bullet points at the end of each chapter. Exercises are also included at the end of each chapter as one mechanism for thinking critically about some of the issues raised about young people, place and identity. I also include a list of 'suggested further reading' for each chapter in order to flag up key journal articles, chapters of books, books or reports that develop further some of the themes raised.

Text boxes throughout the book offer summaries of research about a particular issue in young people's lives (e.g. leaving home) or focus upon a particular group of young people (e.g. Muslim youth). Appendix A provides background information about some of the researchers who have made significant contributions to the field, including information about some of their key publications. This list of authors is not meant to be exclusive or hierarchical and there are many influential youth researchers who are not included. Instead, the 'key authors' have been chosen to represent the diversity of research about young people, place and identity. In the appendices, I have also included a list of research centres and journals that publish work about young people, place and identity.

This book speaks only to the experience of certain groups of young people. Indeed the majority of the examples used, and the focus of this discussion, is young people living in 'Western' societies. This is largely because issues relating to children and young people living in the developing world are often very different in a whole host of ways (see Ansell 2005 for an excellent account of children, youth and development issues). However, this is not to say that all young people in the developed world experience their youthful situations in similar ways. Instead, as I hope this book displays, young people's lives are extremely diverse, rich and often complicated, influenced as they are by a range of different constructions of place, identity and age.

In terms of the overall structure of the book, the majority of the chapters focus upon scales – such as the body, home or nation – or sites and themes – such as public space or migration – in order to demonstrate the ways in which the particular scale, theme or site being discussed shapes and is shaped by young people, place and identity. Overall, however, these different scales and sites overlap, intermingle and connect in diverse ways, and so each chapter should not necessarily be regarded as one narrative; instead I hope that relations and associations can be seen between the ways in which different scales, sites and themes interconnect, influence each other and work in different ways in young people's lives. Before considering how different scales and sites relate to young people, place and identity, Chapters 2 and 3 explore some of the issues involved in researching young people, including considerations relating to research methods, the research process and ethical issues. I have put these chapters before the scales, sites and themes, as a number of the issues raised set the context for the remainder of the book. When assessing research about young people, place and identity, it is always useful to be able to evaluate the quality of the research that has been conducted and this requires an understanding of the nature of research, choices made with regard to research methods and processes, and how ethical and political issues in research are handled by researchers.

Key themes

- Although normally applied to people aged 16–25, definitions of what it means to be a young person are very diverse as age can be viewed as a chronological, physiological or social category.
- Identities are about similarities and differences, and rely on processes and relationalities to give them power and meaning.
- Place can be regarded as a physical location that has connections and relations with other places, as well as an identity that relies on identification with it (or not) to increasing its saliency.
- Lifecourse, intergenerationality and intersectionality are three concepts that can be applied to young people's lives in order to help understand their relationships with place and with their other social identities.
- Youth research is informed by a diverse range of approaches which are interconnected with various disciplinary approaches and responses to changes in society and the economy.

Project ideas

Having read Danny Dorling's account of his journey to school in Box 1.2, think about your journey to school when you were 5, 10 and 15, or your route to school, university or work when aged 20. What was your journey like? Why did you take the route you did and avoid others? Were there particular streets or locations you avoided and why? How has your everyday mobility changed over time?

Thinking about your experiences as a young person, how do concepts such as the lifecourse, intergenerationality and intersectionality map onto your experiences?

Choose a selection of newspapers from the last couple of weeks and select those articles that focus on young people. What discourses of youth are being presented in the articles? What stereotypes of young people are being reinforced?

Suggested further reading

This article offers an account of the importance of place and the influence of neighbourhood spaces:

Dorling, Danny (2001) Anecdote is the singular of data. *Environment and Planning A* 33 1335–69.

This insightful book details the historical development of the concept of 'youth' and how this is related to broader changes in politics and the economy:

France, Alan (2007) *Understanding youth in late modernity*. Maidenhead: Open University Press.

An overview of relational geographies of age, this article maps out three approaches to understanding age: intersectionality, intergenerationality and lifecourse:

Hopkins, Peter and Pain, Rachel (2007) Geographies of age: thinking relationally. *Area* 39(3) 287–94.

This book provides a valuable series of insights into understandings of identity:

Lawler, Steph (2008) *Identity: sociological perspectives*. Cambridge: Polity.

Part I

Researching young people: methods and ethics

2 Research with young people

Understandings about young people's identities and their engagements with different places rely on the findings of research that is conducted by academic researchers – in a range of disciplines – as well as by researchers working in government departments, research consultancies or the voluntary sector. In order to engage critically with research about young people, place and identity, it is important to have an understanding of the methodological and ethical issues that arise in such work. This is particularly important because, although there is much scholarship that provides engaging insights into young people's lives, there is also research that lacks attention to young people's experiences or is tokenistic with regard to young people's voices and perspectives. By appreciating the complexities of the different methods and methodological issues arising in research with young people, you will be better able to value critical and engaged research as well as understand the limitations of research that lacks attention to the details of young people's lives or misrepresents their experiences and circumstances.

Mark Cieslik (2003: 1) notes that 'curiously, although young people have been at the centre of much research, policy-making and practice . . . there have been very few texts exploring the methodological issues facing youth researchers.' Heath *et al.* (2009: 1–2) also point out that, despite the large volume of research about young people, 'there are surprisingly few current textbooks which focus exclusively on the specific methodological challenges of conducting youth research'. This contrasts with understandings about conducting research with children where there is a broad range of textbooks, edited collections and general guidance for researchers (e.g. Greig *et al.* 2007, Greene and Hogan 2005, Kellett 2005). Some of these texts tend to conflate the complexities of youth research with those of childhood studies, as observed by Heath *et al.* (2009). Whilst there are some common methodological issues arising within both traditions, I agree with Heath *et al.* (2009) in their insistence that there are distinctive aspects to studying young people's lives. Youth research has 'distinct histories, theoretical perspectives, methods and key literatures' (Heath *et al.* 2009: 2), and young people also occupy particular positions within society – positions which are different from both adults and children.

As any researcher working with young people will confirm, researching young people involves important decisions relating to the overall philosophical and methodological approach to take, the methods to be used, how the research should be disseminated and many other questions relating to the research content and process. These decisions are mediated by a series of moral and ethical issues that I explore in more depth in Chapter 3. In this chapter, I focus specifically upon decisions about young people's place in research, research methods as well as data analysis and dissemination, in order to explore some of the issues involved in conducting research with young people. These issues should be read in conjunction with the ethical considerations explored in Chapter 3.

Young people's place in research

The ways in which researchers see young people's place in research often reflect beliefs and assumptions they have about the place of young people in social life more broadly. Alongside the various disciplinary and theoretical perspectives discussed in the previous chapter, it is important to reflect upon the place that young people have in research and the extent to which researchers involve young people in certain stages of research or throughout the whole process. How young people are regarded by a researcher influences the philosophical approaches, methods and processes that a researcher is likely to adopt. Sara Kindon (2005) provides a useful account of the ways in which the attitudes of a researcher relate to the relationships they have with those they work with (see Table 2.1). Applying this to researching young people demonstrates that the level of participation of young

Table 2.1 ***Attitudes to relationships between researcher and young people***

Attitude of the researcher and example of this attitude reflected in what a researcher might say to the young people being researched	*Relationship between researcher and young people*	*Relationship between research and youth*
Elitist 'Trust and leave it to me. I know best'	Researcher designs and carries out research; young people chosen but largely uninvolved; no real power sharing	ON
Patronising 'Work with me. I know how to help'	Researcher decides on agenda and directs the research; tasks are assigned to young people with incentives; no real power sharing	ON/FOR
Well meaning 'Tell me what you think, then I'll analyse the information and give you recommendations'	Researcher seeks opinions from young people, but then analyses and decides on the best course of action independently; limited power sharing	FOR/WITH
Respectful 'What is important to you in research? How about we do it together? Here's my suggestion about how we might go about this'	Researcher and young people determine priorities, but responsibility rests with the researcher to direct the process; some power sharing	WITH
Facilitative 'What does this mean for you? How might we do research together? How can I support you to change your situation?'	Researcher and young people share knowledge, create new understandings, and work together to form action plans; power sharing	WITH/BY
Hands-off 'Let me know if and how you need me'	Young people set their own agenda and carry it out with or without the researcher: some power sharing	BY

Adapted from Kindon (2005: 209)

people in research is likely to vary according to the extent to which a researcher's attitude is characterised by either a patronising or elitist tone or a facilitative, respectful and hands-off approach. For example, there are many adults who are used to having power over young people – perhaps in an abusive manner – and so a researcher who holds these values may work in a patronising and elitist fashion in an attempt to control and manage the research participants. In this case, the researcher is working *on* or *for* young people, rather than working *with* them. Furthermore, it is important to be critical about the language used in research because some researchers may discuss working *on* young people but will be engaging with their research participants in an open and engaging manner, whereas some researchers use language that is facilitative and respectful yet may be working with young people in ways that lack sensitivity to the feelings and opinions of the young people involved in their research.

Closely related to this – and reflecting upon work with children – Samantha Punch has queried the extent to which research with children is the same as, or different from, research with adults. She observed that there has been a tendency to see children as either the same as adults or completely different from adults, and noted that it is contradictory that some of those working within the new sociology of childhood often call for the use of innovative research techniques with children, yet are also emphasising the competencies of children. If children are competent social actors, why – asks Samantha Punch (2002: 321) – are special 'child-friendly' methods needed to communicate with them? So, 'the way in which a researcher perceives the status of children influences the choice of methods' (Punch 2002: 322). Researchers who see young people as 'essentially indistinguishable from adults' (James *et al.* 1998: 31) will tend to use the same methods as they use when doing research with adults as they regard young people and adults as basically being very similar. So, if a researcher sees young people as viable and competent social agents then they are likely to treat them as such in research and adopt philosophies, methods and approaches reflective of this. They may choose to employ approaches that actively involve young people in the research process in facilitative and respectful ways (see Table 2.1), and may include young people's voices and positive representations of young people in their work. Contrary to this, a researcher may see young people as being entirely different from adults and so will use methods that reflect this sentiment.

However, Samantha Punch (2002) notes that another approach can be identified which sees young people and adults as having much in common, whilst also recognising that young people may have different competencies. With this approach, researchers who might use traditional 'adult' research methods when researching young people may change them slightly in order to attempt to make them more youth-centred, or they may take a hands-off or facilitative approach by working alongside young people who are themselves conducting research that is important to them. In attempting to explain why differences exist between younger people and adults in research, Punch identifies three main factors. First – and as mentioned in the introduction – young people are constructed and institutionalised in a particular way within society and so there are complex relations of power that influence the relationships between young people and adults in research. Second, there are adults' perceptions about young people which may include fears, assumptions or stereotypes about the expected behaviours and attitudes of young people, including the role they should have in research. Third, young people are different from adults in certain ways. Unlike young children, young people might not have limited or different vocabulary from adults, often are not necessarily physically smaller than adults and may have as much life experience as many adults. However, for the majority of cases, it is likely that young people will be different in important ways from adults whether this is due to their physicality, language, experience, confidence or other factors. Again, this highlights the importance of considering the relationality of being a young person. These three approaches are mapped out in Table 2.2.

Table 2.2 ***What is different in research with young people and why?***

Research issue	*What is different?*	*Why?*
Not to impose researcher's own perceptions	As adults we have all been (or perhaps still are) young people, we think we know about youth, but we see the world and our own experiences of being a young person from an adult perspective	(2) Adult – danger of imposing adult views because of our assumptions about young people (3) Young person may have a different way of viewing the world
Issues of validity/ reliability: subjects may exaggerate or lie to please the researcher	Young people are potentially vulnerable to unequal power relationships in research	(1) Young people are used to having to try to please adults, and may fear adult reactions (2) Adults are used to controlling young people and, in some cases, abusing their power
Clarity of language	More conscious use of language	(2) Adult perceptions of young people's lack of articulateness (3) Young people may have limited vocabulary and use different language
Research context and setting	Many research environments are adult spaces where young people have less control	(1) Youth – adult spaces dominate in society so it can be difficult to find youth spaces in which to conduct research (2) Adults assume that young people would prefer their own spaces
Building rapport	Adults may lack experience in building rapport with young people	(1) Youth – young people's status in adult society means that researchers have to build rapport not only with young people but also with gatekeepers (2) Adult – fears of not being patronising, behaving appropriately, and finding common ground but not faking
Analysis: care not to impose inappropriate interpretations	Ultimately the power lies with the adult researcher to interpret young people's perspectives	(1) Youth – young people's generational position tends to mean that an adult has access to wider knowledge and is better able to analyse young people's social status

(*Continued overleaf*)

Table 2.2 *Continued*

Research issue	*What is different?*	*Why?*
		(2) Adult – danger of imposing adult interpretations because of our assumptions about young people (3) Young people may not fully understand the adult world
Using appropriate research methods: attempts to use the research subjects' preferred methods, and familiar sources of techniques	More attempts to make research with young people interesting and to tap into their interests: for example, photographs or drawings	(1) Youth – young people tend to lack experience of adults treating them as equal and may lack confidence in a one-to-one situation and unfamiliar adults (2) Adults presume that young people prefer these methods, are more competent at them, and that they have a shorter attention span (3) Young people are more used to visual and written techniques at school

Notes: (1) Young people constrained by adult society, (2) Adult perceptions of young people as different, (3) Young people are different from adults

Adapted from Punch (2002: 326–7)

A researcher's perspective on young people's place in research and the focus of the topic being researched are both likely to influence the ways in which research projects are designed. Research design is often overlooked as an important consideration in discussions about ethical research with young people (Hopkins 2008b), yet it clearly has important implications for how young people are accessed and involved in research. Alderson and Morrow (2004) discuss the importance of the selection of young people and the framing of the project as important considerations in research design. Additional issues here include 'the construction and particular ordering of key aspects of the overall procedures such as interview schedules or other materials used during the research' (Hopkins 2008b: 39).

There is also a range of important methodological considerations when doing research with young people and these vary according to the methods being used, the particular groups of young people involved and the topic being researched. There is not enough space to explore the complexities of these issues in this chapter although the chapter which follows explores specific ethical issues in more depth. However, an important factor in research about young people, place and identity is to consider the importance of place or the influence of context on the research interactions. Where research takes place often has significant implications for the nature of research interactions, the type of data collected, and the comfort of the young people involved. Heath *et al.* (2009) point out that place is rarely neutral and is instead imbued with relations of power, privilege and control. Furthermore, particular locations may place the researcher or researched at risk, and so it is often advised – particularly with interviews – to conduct interactions in a public place (Heath *et al.* 2009). That being said, as Lincoln (2005) found in her research about teenage bedroom culture, the focus of the research may result in the opposite being the case.

As Elwood and Martin (2000: 649) suggest:

> the interview site itself produces 'micro-geographies' of spatial relations and meaning, where multiple scales of social relations intersect in the research interview. Careful observation and analysis of the people, activities, and interactions that constitute these spaces, of the choices that different participants make about interview sites and of participants' varying positions, roles and identities in different sites can illustrate the social geographies of a place.

As such, where interviews (or indeed, other forms of data collection) take place offers insights about place and how place is used and understood by young people. These observations are therefore a source of data.

The artefacts, decorations or posters in different places may signify certain things about young people's identities and practices and also locate young people in the context of the social relations of which they are a part and so help us to understand their multiple identities.

It is also useful to reflect on what young people might gain from being involved in research. Malcolm Hill (2006) points out that several research projects have highlighted that children are often likely to be enthusiastic and willing to take part in research whilst also being self-protective, tending to provide short answers as well as being subversive (such as through providing joking answers). Furthermore, Hill (2006) demonstrates that there are four discourses that might influence young people's willingness to take part in research. First, there is young people's interest in the topic being researched. Second, there is the extent to which young people might learn from participating in research. Third, taking part in research may be therapeutic and may offer space to young people to talk about problems they are experiencing. Fourth, research may be a source of empowerment for young people as researchers listen to what they say. Box 2.1 summarises some of the implications that

Box 2.1

Some implications for (adult) researchers from what children have said

Fairness	Ensure as many types of young person and viewpoints as possible are included.
Effectiveness	Try to ensure that the research or consultation will benefit young people.
Agency	Benefit from people's ideas about the best ways to explore their worlds.
Choice	Maximise the opportunities for participants to choose forms of communication and levels of involvement they prefer.
Openness	Be clear to young people about limitations to their participation and the effects it will have.
Diversity	Use a range of methods and include all major perspectives.
Satisfaction	Make the experience a comfortable one and, when appropriate, good fun.
Respect	Recognise young people's rights and opinions; minimise use of power.

Adapted from Hill (2006: 85)

Malcolm Hill (2006) found with regard to what young people said about being consulted.

Research methods

A researcher's attitude about young people as well as the ways in which research with young people might be different from research with adults should be considered alongside issues and decisions about research methods. I draw attention to the methods that researchers tend to employ when researching young people, place and identity, in order that the methodological issues discussed later have some grounding. My focus here tends to be on qualitative research which aims to explore the feelings, meanings and experiences of young people although I also discuss the use of survey data which can be used quantitatively. I discuss these different approaches individually, but there is much to be said for combining different methods in order to explore particular issues using different techniques. It can be useful to combine quantitative and qualitative research in order to contextualise young people's experiences in social trends for which statistical data might contribute insight to. Here, I focus upon the four main sets of approaches used in research about young people, place and identity: verbal, observational, visual and textual, and participatory.

Verbal approaches

Some of the most common methods used in research with young people are those that rely on verbal communication either individually or in group settings. There is a broad range of approaches within qualitative research with young people that adopt a primarily verbal style, the most common of which are individual and group interviews. Essentially, individual interviews are verbal exchanges where the interviewer attempts to elicit information from a young person. 'The qualitative interview is probably the most widely used research method in youth research' (Heath *et al.* 2009: 80). Interviews can be useful for accessing deep understandings and experiences, exploring complex behaviours and motivations and, through being individual, they give priority to individual young people's experiences. They can also be useful in helping young people to reflect upon their circumstances, raise issues with the researcher (that were perhaps not considered by the researcher or their team) and find out more about research. Heath *et al.* (2009: 80) note that qualitative interviewing is grounded in an interpretivist epistemology which means that it is an approach 'which emphasises the subjective meaning of social action, and therefore gives priority to seeing the world through the eyes of those who are being researched'. One of the principal motivations for using interviews is therefore to allow young people to talk about their experiences in their own words, and so it is seen as a youth-friendly approach. Furthermore, given that young people are often misrepresented by the media and regarded as a threat to the moral order of civilised society, interviews are often seen as an important method whereby young people can speak and be heard.

Interviews can vary widely depending on the precise purpose for which they are being used, as well as the style of the researcher. Interviews with young people often have a semi-structured format where the researcher has a set of themes or questions which are used to structure the discussion. Alternatively, some researchers prefer to use unstructured interviews where young people are free to construct their own narratives without being guided by particular questions. Moreover, whether very tightly or loosely structured, interviews can also vary in their level of focus. For example, some researchers may focus on a particular topic, such as leaving school, and so all of their questions might be tightly focused and structured around this topic. Alternatively, adopting a more open approach, they may explore this topic in dialogue with a young person, raising connected issues or points of relevance to the topic. As such, there are different interviewing strategies available to youth researchers, each of which has its own specific strengths and weaknesses which may be more or less prevalent in different places and at different times. Some of the advantages and disadvantages of using interviews with young people include:

Advantages

- Rich and detailed data can be collected about each individual young person's opinions and experiences.
- Young people may value the privacy of individual interviews, especially where the topic is sensitive.
- The one-to-one setting means that the researcher can focus on the needs of the individual young person, trying out a range of creative techniques to see what works. Likewise, in this setting the researcher can be especially attentive to subtle cues suggesting discomfort or a desire to end the interview.

Disadvantages

- Young people may be uncomfortable with the one-to-one setting, particularly if they have had negative experiences of interviews with teachers, police, social workers and so on. Similarly, adults may feel anxious about child protection issues in a private setting. Two alternatives would be to offer interviews with friendship pairs, or to recruit young people to interview their peers. Both solutions have drawbacks, especially where the topic is sensitive.
- Short contact time may limit possibilities for follow-up work.
- Richness and diversity of data can make analysis challenging.

(Adapted from Gallagher 2009a: 75)

Semi-structured interviews tend to rely on the use of an interview guide which contains a set of themes or questions to structure the discussion (see Box 2.2). It is important to note that an interview guide tends not to be used as a fixed set of questions and is instead used in an open and flexible manner. As such, some questions might be more or less relevant depending upon the particular experiences and circumstances

Box 2.2

Interview guide

Your current situation

- Could you tell me about your current situation?
- Are you studying at the moment? (school/college)
- What do you do in your spare time? (TV/sport/other)
- How are you finding your accommodation?
- Who are the main adults that you have contact with? (teacher, support worker, social worker)
- Do you feel that you are given enough support from these people?
- Would you like to have a designated guardian mentor?
- What are the main services that you have contact with? (social work, housing, youth work, etc.)
    - How do you feel about the adequacy of these services? (gaps, strengths and weaknesses, best and worst services)
    - Is there a particular service that you have been unable to find/access?
- Do you feel that all of your needs are being met? (support, counselling, health, education, legal)
- Is there a particular need that you think is more neglected than others?
- What is your current asylum status?
    - How do you feel about that?
- What do you hope to do in the future?

Excerpt from an interview schedule used by Hopkins and Hill (2006) in their research with unaccompanied asylum-seeking children.

of the young person being interviewed and so the interview schedule will be adapted where appropriate during the discussion. Furthermore, adopting a semi-structured approach allows young people to explore other issues that may not have been raised by the researcher. As Heath *et al.* (2009) note, the tone and direction of questions are also important in order to avoid young people giving the 'right' answer that they may be accustomed to saying to adults in authority. It is also important to use open questions that allow young people to explain their answer rather than simply saying yes or no.

Recently, adopting an unstructured approach to interviews with young people has become increasingly popular within youth research. This approach tends to involve a process 'whereby participants are invited to reflect upon specific events in their lives through telling stories about them, or through being invited to reflect upon a particular period in their lives, sometimes in relation to an identified theme or themes' (Heath *et al.* 2009: 83). Such an approach often uses the narratives constructed by young people in interviews as a mechanism for exploring the complexities of young people's social worlds. It has been identified as being particularly useful in understanding young people's biographies alongside the 'critical moments' in their transitions to adulthood (Thomson 2007).

Group discussions also rely on adopting a verbal approach, except this time the discussion takes place with several young people rather than individually. There are a number of different ways of using group discussions. One approach might be in-depth and could involve the group meeting over a number of weeks or months, having related discussions at each meeting. A more commonly used approach is groups that meet only once to discuss a particular topic, often called a focus group. Group discussions can be useful in that they can help to develop ideas, challenge themes and find out about the contested issues in young people's lives through the dialogue created in the group. Some young people may be more comfortable speaking in a group setting and regard it as less threatening than an individual interview. However, group discussions can also be hard to recruit to, challenging to convene, and certain types of young people may dominate the discussion. Overall, there are complex sets of issues that might influence the nature of discussion (see Box 2.3), including the age of the young people, their social composition, the location of the discussion, the sensitivity of the topic and the ethnographic potential of the setting (Allen 2006, Hopkins 2007c, Hyams 2004, Hyde *et al.* 2005).

Some of the advantages and disadvantages of using verbal group-based research methods include the following:

Advantages

- Many young people enjoy being with friends and feel more comfortable when they outnumber adult researchers.
- Group activities can be used to make the research more enjoyable and to cater for diversity.
- Young people may be in pre-existing groups that can be used: friendship groups, youth groups and so on.
- In projects with overt political aims, working with groups can promote an ethic of co-operation and mutual aid, helping the young people to cement their relationships, identify shared goals and spur each other into action.

Disadvantages

- Group dynamics may cause problems, for example one or two young people may dominate; young men may dominate young women (or vice versa); young people who are not friends may not trust each other due to past experiences; shy members may feel unable to voice their opinions in front of the group. Pre-existing power dynamics between peers can be complex and difficult to 'suss out', let alone challenge, without more long-term work.
- Lack of privacy makes discussion of sensitive topics problematic, though role-plays and vignettes can be used to de-personalise.
- Groups can drift off topic easily.
- It may be seen by commissioning bodies as a cheap way to obtain the views of large groups of young people.

(Adapted from Gallagher 2009a: 77)

Although I have discussed individual and group interviews separately here, there is much to be said for using them together in work with

Box 2.3

Some considerations for focus group work with young people

The number of young people
The social composition of the young people
Location, timing and context
Sensitivity of the topic
Age of the participants
Positionalities of the researcher and young people involved*

The ethnographic potential within group interviews
The potential of focus groups to capture group members' vulnerabilities
Challenging dominant group views though group member invigilation
The formation and composition of focus groups
The potential and pitfalls of information exchange in focus groups
The possibility of triangulating qualitative focus group data with a post-interview quantitative check on the validity of the data gathered†

* From Hopkins (2007c)
† From Hyde *et al.* (2005)

young people as some of their specific benefits are complementary. Group discussions can be used as a basis for building up trust with young people who may later take part in an individual interview, or some young people may prefer to discuss issues in a group setting rather than individually, or vice versa. Clearly, with both individual and group interviews there is a range of methodological considerations about access, consent, power relations and so forth, and I explore these issues in greater depth later in this chapter and in the chapter that follows.

Observational approaches

Ethnographic and observational approaches to researching young people were used frequently by many of the members of the Centre for Cultural Studies at Birmingham University and are still employed regularly by researchers aiming to explore young people's lives. As Heath *et al.* (2009: 99) note:

> Ethnography has long held a special place within youth research. Some of the classic ethnographies conducted by pioneers from the Chicago School from the 1920s onwards, such as Thrasher's *The Gang* (1927) and Foote Whyte's *Street Corner Society* (1943), have a central focus on young people's experiences. Even today, 30 years on from its publication in 1977, Willis's *Learning to Labour* remains one of the most well-known ethnographies not just within youth studies but within British sociology more generally.

Ethnography normally refers to a form of data collection that relies on observing people in their ordinary settings and so is seen to give access to young people's everyday activities and the social values and meanings connected with these. As such, it often requires the researcher to spend lengthy periods of time with the young people they are researching (Edmond 2005). Although ethnography relies largely on observing young people within a particular setting, participation is also an important element. So, an ethnographic study of a youth club will involve lots of observation, although the researcher may also participate more actively within the activities of the club, rather than being a passive observer. Often, the most important data created by the ethnographer are sets of fieldnotes containing details of their various observations about the happenings in the location being studied. Many researchers who adopt observational approaches also use verbal techniques, and so may make ethnographic notes whilst interviewing young people individually at different points over the course of a research project. Some of the advantages and disadvantages of ethnographic work with young people include:

Advantages

- Can be used to explore what young people do, as well as what they say.
- Can be inclusive, as the focus on developing relationships with young people enables flexibility around communication styles.
- Less disruptive to young people's everyday activities than other methods. Observation can 'fit in' with whatever they are doing – work, hanging out and so on.
- Participant observers are usually able to assist participants in various ways with their daily activities such as helping with school work or assisting with the running of community groups as a volunteer. Ethically, this may be seen as an advantage over more extractive, less interactive methods.

Disadvantages

- Can be perceived as intrusive by young people or their adult caregivers.
- Informed consent is usually difficult to negotiate.
- In participant observation, the development of strong relationships can create complications: for example, allegiance to rival groups of young people; observing illegal or forbidden activities; negotiating the end of fieldwork.

(Adapted from Gallagher 2009a: 78)

An excellent example of research that uses observational approaches to understand young people, place and identity is the ethnographic work conducted by Anoop Nayak (2003a, b, 2006) in the north-east of England. In his account of the intersections of youth, gender, race and place, Nayak used observational approaches alongside individual interviews. He recorded his observations in a fieldwork journal that enabled him to record events as they happened. 'This meant that an in-depth insight into individual social life-paths could be garnered over time and participants could be observed in different places and situations' (Nayak 2003a: 6).

Visual and textual approaches

A third set of methods used in seeking to understand young people's lives relies upon approaches which focus on the use of visual and textual materials alongside representational aspects of youth. Many young people are now bombarded by visual images whether this is through the internet, on their mobile phones or through images downloaded from a digital camera. Alongside this, researchers are increasingly using methods and approaches with young people that rely on the creation of, or use of, visual materials. In particular, the use of photography, video and mapping are now common features of much research with young people. Heath *et al.* (2009) usefully distinguish between three approaches to the use of visual materials in youth research. The first involves the analysis of visual materials that exist already in magazines, websites or on television. The second focuses on visual materials produced by the researcher such as photographs of particular places. Third, there are visual materials that are generated by young people themselves. In their work with young Muslim-American women, Mayida Zaal *et al.* (2007) used material generated by their young participants in their analysis. They asked their participants to draw maps of their social identities and then analysed these alongside their other data to explore the multiple identities of the young women.

There is also a range of textual sources that researchers might use to access the social worlds of young people. These include books, magazines or albums about young people, websites, newspapers or song lyrics to name a few. Like visual materials, textual sources may include those already existing or may involve either the researcher or young people creating text for use in the research. Often used in collaboration with other methods, some youth researchers use diaries as a mechanism for accessing the social lives of young people (Barker and Weller 2003). Textual materials tend to be used qualitatively with attention being given to the meanings, discourses and representations evident within the text. However, textual sources can also be used in a quantitative manner whereby the analysis focuses on seeking to give a representative picture of young people's experiences or identities. One example here is where researchers used content analysis to understand young people, place and identity. Sameer Hinduja and Justin Patchin's (2008) work about young people's use of the internet involved a random selection of 9282 MySpace profile pages. They found that 26 per cent of these were created by young people. Moreover, through their detailed analysis, they found that young women created 54 per cent of the profiles and that 57 per cent had at least one picture of a young person, and that more than 8 per cent of the total sample had inflated their age; 40 per cent included their name, 81 per cent their city and 28 per cent their school.

Participatory approaches

Increasingly common approaches used in research with young people are those that draw upon the philosophy and practices of participatory action research. Participatory action research is sometimes conflated with ethnographic approaches characterised by observing people (Winchester 2005) or even qualitative research in general (Greig *et al.* 2007). However, these assumptions are very problematic because – although participatory action research is broadly qualitative – it is also distinct from other qualitative approaches in two important ways. First, participatory action research is underpinned by a rather different set of philosophical assumptions about the nature and purpose of research because its primary focus is often about social change and action rather than just social analysis (Kindon 2005). Second, participatory action research is also different from other research approaches in that it is about working collaboratively *with* young people to achieve change rather than working *on* them (Kindon 2005). As such, it stands as a serious epistemological challenge to mainstream traditions of research (Kindon *et al.* 2007) – some of which are discussed above – and so is really very different from them. That being said, 'participatory action research is a contested concept applied to a variety of research approaches employed in a variety of fields and settings' (Kemmis and McTaggart 2000: 567) and so the work of participatory action researchers working with young people is characterised by incredible diversity. Figure 2.1 highlights some of the techniques that are commonly used in participatory action research.

The unique qualities of participatory approaches mean that they nearly always involve working collaboratively with a group, and sometimes a group of young people may decide to research a particular topic and then ask a university researcher to work with them in the process. On other occasions, the researcher may devise the focus of the research in discussion with a group of young people, or with an organisation or agency working with young people. With participatory research then, the researcher often has far less control over the focus of the research and how the research is conducted as this is likely to be

Figure 2.1 Common methods used in participatory action research.

Source: Kindon, Pain and Kesby 2007: 17

shaped by the young people or other groups involved. Some researchers may also be inclined to include participatory techniques in their research with young people assuming that such approaches may be more interesting and engaging for the young people involved. Although this may well be the case, it is important to ensure that such approaches are not employed in a manner that may appear condescending to the young people involved. Some of the advantages and disadvantages of using a participatory approach include:

Advantages

- Young people may find these methods appealing and enjoyable.
- By engaging with the visual, audible, kinaesthetic and performative aspects of young people's lives, these methods may be inclusive for young people who do not respond well to the more traditional methods of reading, writing and talking.
- Data collected through participatory methods may be more

effective for feedback and dissemination than traditional text-based outputs.

Disadvantages

- More elaborate methods such as video-making or recording a rap may be resource-intensive, requiring time, money, equipment and expertise.
- Data produced may raise problems for analysis. Social researchers have well-established techniques for analysing numbers and texts, but these are not easily adapted to music, video, drama or dance.
- Risk of relying on and reinforcing unhelpful stereotypes about young people (for example, that all young people like drawing)

(Adapted from Gallagher 2009a: 79)

An excellent example of a participatory action research project is the work that Caitlin Cahill (2004, 2007a, b, c, d, e) conducted with young women in New York City (Image 2.1). Here, six young women were recruited through school and community centres to participate in a

Image 2.1 Caitlin Cahill and her young co-researchers.

project in which they were 'involved in every aspect of developing and creating the project' (Cahill 2004: 275). This project took a participatory approach because it 'takes lived experience as the starting point for investigation, places emphasis upon the research process, and reconsiders the value of research as a vehicle for social change'. In the end, the young women were collaboratively involved in the entire research process, including designing a sticker campaign to challenge stereotypes, publishing a collaborative report as well as other research publications and designing a website (see Cahill 2004, 2006, 2007a, b, c, d, e).

Survey approaches

Although the approaches reviewed so far are often useful for understanding the complexities of young people's lives, it can also be beneficial to analyse broader trends across larger groups of young people or between different groups of young people. Using survey data is a useful mechanism for accessing such trends. Gallagher (2009a) observes that surveys and questionnaires are often used in psychological and medical research as well as by social researchers. Youth researchers can use already existing survey data to learn more about young people or they can design their own surveys to find out more about young people, place and identity. Some of the surveys available to youth researchers include national surveys such as census data or other national evaluations about various aspects of social life. Some countries occasionally conduct national youth surveys (e.g. Australian youth survey) some of which are longitudinal in approach (e.g. Longitudinal Study of Young People in England) or have general surveys which may contain data relevant to young people's lives.

Alongside these already existing surveys, youth researchers may choose to conduct a survey of their own. Important considerations here include the survey or questionnaire design, the piloting of this and gaining access to a sample of young people to complete the survey. Plows and Gallagher (2009) discuss using a survey approach in evaluation research of a youth counselling service. They sought to recruit two samples, one being the entire population of a local school and the second being young people who used a local community centre. As the design of a survey is one of the most important considerations when doing this type of research, Plows and Gallagher (2009) were careful to be clear and simple when constructing the questionnaire and redrafted it several times after it was piloted with a smaller group of young people. They also ensured that it was only two pages long and used closed questions. In the end, they accessed the views of 444 young people. Overall then, some of the advantages and disadvantages of using surveys or questionnaires include:

Advantages

- Enables collection of large amounts of data in standardised formats.
- A high level of anonymity can easily be achieved.
- Can be useful to obtain views of young people who would not have the confidence to speak in an interview or focus group.
- Young people may be familiar with the format (for example, those who have experienced written exams; those who have carried out their own surveys as part of school, college or university projects).
- When administered through schools, it is often possible to obtain a high response rate.

Disadvantages

- Young people may perceive them negatively, for example as a piece of coursework; as in intrusion into their private lives; as a boring exercise.
- Voluntary consent is especially problematic for this method. When administered through schools, often young people do not consider non-participation as an option.
- If using self-complete questionnaires, these may exclude young people with low literacy (though researchers can complete these with young people, one to one).
- Design, piloting, administering and data input can be time-consuming and tedious.
- Can produce unwieldy, messy data sets, especially if design is flawed.

(Adapted from Gallagher 2009a: 73–4)

Data analysis and dissemination

In engaging with research about young people, place and identity, it is important to consider the extent to which researchers have given attention to the processes of data analysis and dissemination. The concepts of rigour, validity and reliability are central to the process of data analysis. Rigour refers to the extent to which a study can be classified as being valid and reliable as well as it being conducted in an academically honest, open and reflective manner (Baxter and Eyles 1997). Validity is commonly understood to refer to the 'correctness or credibility of a description, conclusion, explanation, interpretation, or other sort of account' (Maxwell 1996: 87), whilst reliability refers to the consistency or repeatability of outcomes. For many studies, 'the most common ways to ensure rigour are the provision of information on the appropriateness of the methodology, the use of multiple methods, information on respondent selection and the presentation of verbatim quotations' (Baxter and Eyles 1997: 506). As such, a rigorous study with young people is likely to present (anonymised) quotations from young people, details about how the young people were selected as well as clarification of the methodological approach and use of multiple methods. Other strategies include 'details of interview practices, discussions of the procedures for analysis, immersion/ lengthy fieldwork, revisits to respondents, verification by respondents, appeals to interpretive communities and the provision of a rationale for verification (validity) of the findings' (Baxter and Eyles 1997: 508). As mentioned in the discussion about verbal methods above, rigour can also be enhanced through using standardised interview guides alongside being attentive to the relations of power present within interview settings (Baxter and Eyles 1997). Box 2.4 outlines some of

Box 2.4

Strategies for establishing rigour in research with young people

Rationale for methodology
Multiple methods
Information about selection of young people
Present of quotations from young people
Details about interview practices
Analysis procedures
Immersion/lengthy fieldwork
Revisits to young people
Verification of findings by respondents
Appeals to interpretative community
Rationale for verification

Adapted from Baxter and Eyles (1997: 507)

the strategies that may help researchers when conducting research about young people, place and identity.

Jamie Baxter and John Eyles (1997) highlight the importance of credibility, transferability, dependability and confirmability as criteria for evaluating qualitative research. Credibility is about 'the degree to which a description of human experience is such that those having the experience would recognise it immediately and those outside the experience can understand it' (Baxter and Eyles 1997: 512). Transferability refers to the extent to which study findings correspond to contexts outside the research project, and dependability relates to the ways in which the study manages instabilities or changes over time. Confirmability refers to the extent to which the outcomes of research are shaped by respondents' experiences and social worlds rather than by the biases of the researcher (Baxter and Eyles 1997). In a response to Baxter and Eyles (1997), Cathy Bailey, Catherine White and Rachel Pain (1999a, see also Baxter and Eyles 1999, Bailey *et al.* 1999b) developed a set of principles for evaluating qualitative research (see Box 2.5).

In terms of dissemination, it is important that research with young people is disseminated directly to young people in an appropriate format. This may sound obvious, but there is still much research conducted with young people where the research participants are not informed of the outcomes whilst teachers, youth workers or local

Box 2.5

Principles for evaluating qualitative research

The need for theoretical sensitivity
Reflexive management that strengthens qualitative validity
Constant comparison by continued questioning
Thorough documentation of procedures to leave a 'paper trail' audit that strengthens qualitative validity
Clear and open reporting of procedures
Clear discussion of how theory 'fits' the reality of the respondents' lives, with rationale offered for 'negative' cases (not every incidence can fit)
Generation of criteria for evaluation of particular research
Recognition of the researcher(s) influence on the research findings (interpretation is always partial)
Use of archives for data and documentation relating to research procedures

Adapted from Bailey *et al.* 1999a: 175

government officials are. Depending upon the particular group of young people being researched and the sensitivity of the topic, it may be useful to consider holding dissemination events specifically for those involved in the research. Furthermore, depending upon the age of the young people involved in the research, it may be more appropriate for them to receive a youth-friendly report at the end of the study, although there may be occasions where it is appropriate for them to receive the full report and they may indeed feel excluded if they do not receive this.

Key themes

- A researcher's perspectives about the place of young people in research may feed through to the ways in which they involve young people in research and so it is important to be attentive to this when reading research.
- There is some debate about the extent to which children and young people can be consulted using the same techniques and approaches used in research with adults and the extent to which techniques should be adapted according to the age of the participants.
- Much work about young people, place and identity tends to use qualitative methods to explore young people's feelings and values although there is also a tradition of survey-based research within youth research.
- Researchers using qualitative methods tend to mix methods in order to access young people's social worlds from different approaches. Some of the methods commonly used include verbal, ethnographic, participatory and textual and visual approaches.
- Rigorous research about young people, place and identity pays attention to the concepts of validity and reliability, and so seeks to justify the methodological approach, provides (anonymised) information about respondents and includes quotations from young people. Additional attention to specific research practices and processes of verification may also improve the quality of the research.
- It may be useful to produce a youth-centred version of research outcomes for the young people involved in the research although young people may also appreciate the full report. It can be valuable – where appropriate – to include young people in dissemination events or for a separate youth dissemination event to take place.

Project ideas

Select three studies about young people, place and identity and analyse the ways in which the researcher constructs youth and how this is reflected in their methodological approach. Are there any contradictions in their approach?

If you were commissioned to undertake a study about young women's experiences of leaving school, what research methods would you consider employing and why?

Select three studies about young people, place and identity and critically evaluate the rigour, validity and reliability of the research.

Suggested further reading

This useful article charts the ways in which qualitative research can be evaluated in terms of validity, reliability and rigour:

Baxter, Jamie and Eyles, John (1997) Evaluating qualitative research in social geography: establishing 'rigour' in interview analysis. *Transactions of the Institute of British Geographers* 22 505–25.

This brilliant book explores different ways of conducting research with young people:

Heath, Sue, Brooks, Rachel, Cleaver, Elizabeth and Ireland, Eleanor (2009) *Researching young people's lives*. London: Sage.

This useful article reflects upon children's and young people's views on being consulted in research:

Hill, Malcolm (2006) Children's voices on ways of having a voice. *Childhood* 13(1) 69–89.

This collection offers a series of rich insights into the complexities of participatory action research:

Kindon, Sara, Pain, Rachel and Kesby, Mike (eds) (2007) *Participatory action research approaches and methods: connecting people, participation and place*. London: Routledge. 9–18.

Containing a practical set of examples, this book explores issues of ethics, research design and analysis in research about young people:

Tisdall, Kay, Davis, John and Gallagher, Michael (2009) *Researching with children and young people: research design, methods and analysis*. London: Sage.

3 Ethical and methodological considerations

All research with young people involves negotiating an often complex series of ethical and methodological issues. Mark Cieslik notes that there is 'a range of moral and ethical issues confronting individuals who investigate the lives of often vulnerable young people who lack resources, social networks and knowledgeability of those conducting the research' and so researchers need to 'tread carefully and bear in mind how their work may be represented by the media, and how this in turn may impact on the lives of young people themselves' (Cieslik 2003: 1–2). Virginia Morrow (2008: 51) states that 'ethics' can be defined as a 'set of principles and rules of conduct': ethics in research relates to 'the application of a system of moral principles to prevent harming or wronging another, to promote the good, to be respectful, and to be fair'. Moreover, Hugh Matthews (2001: 117) argues that 'ethics is about power – what is recognised as ethical depends on values, moral judgements, perceived goals and intended outcomes.' These issues are central to all research – although especially when working with young people – and should be considered throughout the research process and not just at specific points such as when applying for ethical approval. So, ethical issues should be considered when we are contemplating research, when we are actually doing research and when we are analysing, writing up and disseminating research about young people (Hopkins and Bell 2008). Priscilla Alderson and Virginia Morrow have developed a very useful set of 'ten questions' that researchers should ask themselves when researching children and young people (see Box 3.1). It may therefore be useful to

Box 3.1

Ten topics for consideration in carrying out research with young people

1 The purpose of the research
   - What is the research for (to learn more about young people's views, experiences or abilities, to develop or evaluate a service or product, some other positive purpose)?
   - Whose interests is the research designed to serve?
   - If the research findings are meant to benefit certain young people, who are they and how might they benefit?
   - What questions is the research intended to answer?
   - Why are the questions worth investigating?
   - Has earlier research answered these questions?
   - If so, why are the questions being re-examined?
   - How are the chosen methods best suited to the research purpose?

2 Costs and hoped-for benefits
   - What contributions are young people asked to make to the research such as activities, or responses to be tested, observed or recorded?
   - Might there be risks or costs (time, inconvenience, embarrassment, intrusion of privacy, sense of failure or coercion, fear of admitting anxiety)?
   - Might there be benefits for children who take part in the research? (satisfaction, increased confidence or knowledge, time to talk to an attentive listener?)
   - Are there risks and costs if the research is not carried out?
   - How can the researchers promote possible benefits of their work, and prevent or reduce any risks?
   - How will they respond to young people who wish to refuse or withdraw, or who become distressed?
   - Are the research methods being tested with a pilot group?

3 Privacy and confidentiality
   - How will the names of children be obtained, and will they be told about the source?
   - Will children and parents be able to opt in to the research (such as by returning a card if they wish to volunteer)?
   - Opt out methods (such as asking people to phone to cancel a visit) can be intrusive.
   - Is it reasonable to send reminders, or can this seem coercive?
   - Will research directly with individuals be conducted in a quiet, private place?
   - Can parents be present or absent as the child prefers?

- In rare cases, if researchers think that they must report a child's confidences, such as when they think someone is in danger, will they try to discuss this first with the child?
- Do they warn all children that this might happen?
- Will personal names be changed in records and in reports to hide the child's identity?
- What should researchers do if children prefer to be named in reports?
- Will the research records, notes, tapes, films or videos be kept in lockable storage space?
- Who will have access to these records, and be able to identify the young people? (Using post codes instead of names does not protect anonymity.)
- When significant extracts from interviews are quoted in reports, should researchers first check the quotation and commentary with the child or parent concerned?
- What should researchers do if respondents want the reports to be altered?
- Before researchers spend time alone with young people, are their police records checked?
- Should research records be destroyed when a project is completed, as market and medical researchers are required to do?
- Is it acceptable to contact the same young people again and ask them to take part in another project?

4 Selection of research participants
- Why have the young people concerned been selected to take part in the research?
- Do any of them belong to disadvantaged groups?
- If so, has allowance been made for any extra problems or anxieties they may have?
- Have some young people been excluded because, for example, they have speech or learning difficulties?
- Can the exclusion be justified?
- If the research is about young people, is it acceptable only to include adult subjects?
- Are the research findings intended to be representative or typical of a certain group of young people? (If so, have the young people been sufficiently well selected to support these claims?)
- Do the research design and the planned numbers of young people allow for refusals and withdrawals?
- If too many drop out, the research is wasted and unethical.

5 Funding
- Should the research funds be raised only from agencies that avoid activities that can harm young people?
- Does the funding allow for time and resources to enable researchers to liaise adequately with the young people to collect, collate and analyse the data efficiently and accurately?

(continued overleaf)

- Are the young people's and parents' or carers' expenses repaid?
- Should young people be paid or given some reward after helping with research?

6 Review and revision of the research aims and methods
  - Have young people helped to plan or comment on the research?
  - Has a committee, a small group or an individual reviewed the protocol specifically for its ethical aspects and approach to young people?
  - Is the design in any way unhelpful or unkind to young people?
  - Is there scope for taking account of comments and improving the research design?
  - Are the researchers accountable to anyone to justify their work?
  - What are the agreed methods of dealing with complaints?

7 Information
  - Are the young people and – where appropriate – adults concerned given details about the purpose and nature of the research, the methods and timing, and the possible benefits, harms and outcomes?
  - If the research is about testing two or more services or products are these explained as clearly and as fully as possible?
  - Are the research concepts, such as 'consent', explained clearly?
  - Are young people given a clearly written sheet or leaflet to keep, in their first language?
  - Does a researcher also explain the project and encourage them to ask questions, working with an interpreter if necessary?
  - Does the leaflet give the names and address of the research team?
  - How can young people contact a researcher if they wish to comment, question or complain?

8 Consent
  - Are young people told that they can consent or refuse to take part in the research?
  - Do they know that they can ask questions, perhaps talk to other people, and ask for time before they decide whether to consent?
  - Do they know that if they refuse or withdraw from the research this will not be held against them in any way?
  - How do the researchers help the young people to know these things, and not to feel under pressure to give consent?
  - How do they respect young people who are too shy or upset to express their views freely?
  - Are parents or guardians asked to give consent?
  - What should researchers do if a young person is keen to volunteer but the parents refuse?
  - Is the consent written, oral or implied?
  - If consent is given informally, how do the researchers ensure that each young person's views are expressed and respected?
  - If young people are not asked for their consent, how is this justified?

9 Dissemination
- Does the research design allow enough time to report and publicise the research?
- Do the reports show the balance and range of evidence?
- Will the young people involved be sent short reports of the main findings?
- Will the research be reported in popular as well as academic and practitioner journals, so that the knowledge gained is shared more fairly through society?
- Can conferences or media reports be arranged to increase public information, and so to encourage the public to believe that it is worthwhile to support research?
- Will the researchers meet practitioners to talk to them about practical ways of using the research findings?

10 Impact on young people
- Besides the effects of the research on the young people involved, how might the conclusions affect larger groups of young people?
- What models of youth are assumed in the research (young people as out-of control and rebellious, young people as weak, vulnerable and dependent on adults, as immature, irrational and unreliable, as capable of being mature moral agents, as consumers)?
- How do these models affect the methods of collecting and analysing data?
- Is the research reflective, in that the researchers critically discuss their own prejudices?
- Do they try to draw conclusions from the evidence, or use the data to support their views?
- Do they aim to use positive images in reports, and avoid stigmatising, discriminatory terms?
- Do they try to listen to young people and to report them on young poeple's own terms though aware that young people can only speak in public through channels designed by adults?
- Do they try to balance impartial research with respect for young people's worth and dignity?

Adapted from Alderson and Morrow (2004)

use this chapter and the ten points for consideration in Box 3.1 in collaboration with each chapter rather than seeing this chapter as sitting alone. For this reason, I include occasional exercises about 'ethical and methodological considerations' in some of the chapters that follow.

As Alderson and Morrow (2004) clarify, these questions are not intended to be all encompassing and some may apply more than others depending on the nature of the research being conducted. Not all of the issues discussed apply to all research with young people, as the concerns raised will vary according to the particular group of young people being researched and the context in which this takes place. However, on the whole, researchers are advised to consider these issues in their work with young people and in doing so evaluate and be reflective about the extent to which each issue might relate to their work. In approaching this issue, I take a perspective looking across the research process.

The vast majority of research today – regardless of whether or not it involves young people – requires some sort of ethical approval procedure. This may be an extensive set of forms, committees, panels or meetings or may simply require a one-page tick-box sheet clarifying specific aspects of the overall research project, and such practices vary widely across different national and institutional contexts. There is something of a disjuncture between university and college ethics processes and the negotiation of ethical issues in practice, and these tensions have recently been the subject of much interest, particularly amongst participatory action researchers (Askins 2007, Cahill 2007c, Cahill *et al.* 2007, Elwood 2007). Overall, it is crucial to recognise that, although a project may have received ethical approval from a panel or committee, the negotiation of ethics is a constant process throughout the entire research process and is not simply a one-off event.

Although often not regarded as a particularly sensitive or ethically charged part of the research process, the design of research about young people, place and identity raises significant ethical and methodological considerations. Key questions here include: how young people will be consulted; at what point in the research process this consultation will take place; and what ethical motivations are there behind these decisions about research design. It may be ethically more appropriate to speak to specific groups of young people at particular times, or to ask certain questions towards the end of the research rather than at the start. These research design issues are often overlaid with ethical and methodological considerations. Research about young people is often tokenistic and short-term, focusing on consulting teachers, youth workers or other professionals with the young people either being ignored or consulted quickly at the end of the project. Such research is specifically designed in this way and so those involved are making ethical decisions about the place of young people in research, perhaps without recognising it.

Ethical considerations are also important when negotiating access to young people for research purposes (Valentine 1999). The institutional context, research focus and the sensitivity of the topic being researched are all likely to shape how access is negotiated. Some gatekeepers may resist granting access and this can result in research not taking place and in young people being excluded from having a voice. It may therefore be useful to ask resistant gatekeepers whether they have asked the young people in their care if they are interested in participating. It is also important to be open and honest about the nature of the research, why it is being conducted and who it is for.

Although research which focuses on place and young people is likely to be sensitive to various contextual factors in the research, it is important to be aware of the location, context and setting of the research as this can raise significant ethical and methodological considerations. This may involve being aware of the micro-setting of the research within the room or building in which the project is being conducted as well as considering broader contextual issues. This may include acknowledging factors such as neighbourhood territorialism, institutional pressures or particular interests or practices of the young people involved in the research. Kye Askins (2007: 351) sees 'ethics as emergent through social relations in place.' Furthermore, 'the geographical location where research takes place is important, exerting significant influence over the research process and its outcome' (Hopkins and Bell 2008: 4). We need to think carefully about relations of power, control, inclusion and exclusion and how these work alongside identities in different ways in different places. We should therefore consider the perspectives of the researcher (see Chapter 2), the geographical location of the research and the details of the particular issues being explored.

Upon completing a research project with young people, maximising the impact of such work often relies upon an effective dissemination strategy. Issues to consider here include ensuring that the young people are positively represented in any documents presented and that images of young people are used in a sensitive way. This does not mean that negative or challenging aspects of young people's experiences should not be discussed and explored but rather that researchers should be attentive to how images and representations of young people are used in reports and other outputs. Furthermore, it is regarded as unethical when research participants are not informed of the outcome of research they took part in, and young people are no different. It may therefore be appropriate to invite young people to any dissemination events, or indeed it may be regarded more appropriate to have a separate dissemination event for the young people involved in the research.

Although there are a number of ways in which researchers can conduct ethical research about young people, place and identity, it is important to be aware of practices that are regarded as unethical. Examples of questionable practices may include the involvement of young people in research without their knowledge or consent, or withholding details about the true nature of the research being conducted. It is also unethical to coerce young people to participate in research just as it is unethical to invade young people's privacy or expose young people to any form of stress. Persuading young people to engage in behaviour which lowers their self-esteem as well as displaying general lack of consideration or respect for young people should also be avoided.

Obtaining informed consent

Obtaining informed consent is generally recognised as an ethical practice in much social science research (Heath *et al.* 2009). Furthermore, as Malcolm Hill (2005) notes, young people's right to give or deny consent to participate in research is something they share with adults but also shows respect for who they are. Michael Gallagher (2009a) clarifies that informed consent is generally understood as relying on four key principles. First, informed consent is understood to involve an action such as openly verbally agreeing to participate or by signing a consent form or agreement. Second, informed consent requires research participants to be fully informed about the aims, methods and likely outcomes of research, and so a leaflet explaining the research as well as verbal clarification about the overall research is expected. Third, informed consent should be given voluntarily and should not involve any form of coercion or abuse of power relations. Fourth, informed consent is something that is constantly open to negotiation as young people have the right to drop out of a research project at any time, without giving a reason and may refuse to answer specific questions. Furthermore, given the power inequalities that exist between adult researchers and young people, it may be useful to regularly remind young people that they may drop out of the research at any time and can refuse to answer particular questions without giving a reason. Hill (2005) notes that for consent to be valid, young people should be informed of the aims of the research; the time commitment required; who will be provided with the results; whether or not feedback will be given; and the extent to which participation is confidential. Box 3.2 suggests some of the stages involved in gaining informed consent.

The age of young people involved in research often adds an additional layer of complexity to the obtaining of informed consent. In most contexts, young people who are adults can be expected to openly consent – in an informed way – to participate in research. However, when the young people involved are aged under 18, obtaining informed consent is more complex. Although most teenagers possess the agency and competency to decide whether or not to participate in research, it is often thought to be useful to obtain parental consent alongside the consent of the young person. Alderson and Morrow (2004) observe that the legal positioning with regard to parental consent in the UK is not clear, especially given the recognition that competent children can give valid consent. 'Competence is defined as having 'sufficient understanding and intelligence to understand what is proposed' and 'sufficient discretion to enable [a child] to make a wise choice in his or her own interests' (Alderson and Morrow 2004: 99). In respecting

Box 3.2

Gaining informed consent

- Introduce yourself clearly, stating who you are, where you have come from, and what you are there to do.
- If you have an information leaflet, use this as the basis for discussion. Ask young people if they have any questions.
- Another approach is to ask young people what they think they are there to do. Then fill in the gaps, using your leaflet if need be.
- Emphasise that they don't have to take part, that they don't have to answer a question if they don't want to, that they don't have to say anything if they don't want to. It can be useful to work these principles into a group agreement if you are using one.
- If working in a school, explicitly state that this is not a piece of school work, and that they won't get into trouble if they don't take part
- If you are going to record the interviews, tell young people why and who will listen to the recordings, and then ask them how they feel about this, stressing that it is their decision.

Adapted from Gallagher (2009a: 136)

young people's competencies, it is also important to remember that young people should openly consent to their participation in research and therefore it should not be assumed that because a parent has given consent that a young person definitely wants to participate in research.

Depending upon the sensitivity of the topic and the personal backgrounds of the young people involved in the research, it may be very time consuming to obtain informed consent. A number of stages are often involved including: 'fully explaining the purpose of the research; clarifying the motivations of the researcher; outlining details about the funding of the research; giving time for research participants to read fully the material you provide them with (including permission slips etc.)' (Hopkins 2008b: 41). In some of my own research with unaccompanied asylum-seeking children and young people, I found that the informed consent could take around 30 or 40 minutes to obtain. There were a number of reasons for this. First, many of the young people had been interviewed by immigration officials and social workers and so were concerned about the influence that their participation may have on their asylum applications. I therefore took the time to explain the background and motivations of the research and clarified what was meant by confidentiality. Second, some of the young people did not speak English and were working with interpreters who needed time to translate the information leaflets and consent forms for the young people. Third, many of the young people were apprehensive about signing a consent form and wanted to ask questions about what would happen to the forms after the research had taken place (Hopkins 2008b). This is a very specific example but is useful in highlighting how important it is to think through the ethical issues and time that may be required to obtain informed consent. Issues here may include the sensitivity of the topic, the age, competency and social positioning of the young people as well as their life experiences.

Confidentiality and anonymity

Principles associated with confidentiality and anonymity sit alongside informed consent as key foundations of ethical research. Although these terms are regularly used interchangeably, there are important differences to point out. 'Anonymity refers to the protection of the specific identities of individuals involved within the research process, whereas confidentiality refers to promises not to pass on to others specific details pertaining to a person's life: a "between you and me" sort of approach' (Heath *et al.* 2009: 34). The principle of anonymity is often assumed to be the norm in social research but this can be challenged when young people ask for their actual names – rather than pseudonyms – to be used. This can be a tricky issue to deal with. It is important in such cases that young people are made as fully aware as possible of the consequences and potential repercussions of their name appearing in a final report. This is particularly the case in certain types of research – such as evaluation-based work – where teachers, social workers or youth workers are likely to read the outcomes of research and will see what young people have said about their use of and experience of particular services (Hill 2006). Furthermore, revealing the name of one young person might have the consequence of making other young people's names identifiable. Heath *et al.* (2009: 34) suggest:

> Researchers need then to make judgements about the pros and cons of allowing young people's real names to be used in research outputs, based on their greater knowledge of the likely harm that might be caused to a young person if their real names are indeed used, and should advise them accordingly.

Overall then, the tendency in research with young people is to replace young people's actual names with pseudonyms. It can also be useful to ask young people to give themselves an alternative name so that they have a say in the pseudonym used in place of their name. Gallagher (2009a) also suggests that names could be avoided altogether in order to manage the issue of anonymity.

As well as anonymising the names of young people, some researchers change the name of the localities in which research is conducted as an additional strategy for protecting the identities of the young people involved in the research. Some researchers choose to use pseudonyms for places whereas other researchers might generalise and refer to a city or region rather than the specific neighbourhood or community in which the research was conducted. Nespor (2000) has been critical of assumptions about the need for places to be anonymised for research to be ethical. He argues that changing the names of places where research has taken place does not necessarily protect the identities of individuals. Furthermore, he argues that it is wrong to assume that findings from one place can be transposed to other research settings due to the importance of the context in which research takes place.

This can then be seen to distance young people from the specifics of local social, political and economic issues and the process of anonymising place can limit the extent to which research might inform local practice. Overall, the anonymising of place requires researchers to balance out desires to protect the identities of the young people involved in the research against wishes to inform local practice or have their research clearly embedded in the local context. This is a particularly challenging issue for researchers interested in young people, place and identity given the attention that such work often gives to the influence of the local context and the importance of place.

Confidentiality normally refers to the practice of information (personal details, interview data, etc.) being treated in a private manner and therefore not shared with others without the permission of a young person. This means that information collected by a researcher should be carefully managed and stored in a lockable filing cabinet or in a computer folder that is protected by a password (Gallagher 2009b). There are different levels and forms of confidentiality and times where it can be appropriate to breach confidentiality (see Box 3.3), depending on the particular group of young people being researched and the nature of the research topic. For example, Malcolm Hill (2006) differentiates between public confidentiality – which is about not identifying participants in research reports and presentations – and social network confidentiality – which is about not passing on information to family members, peers or other people known to the young people involved in the research. Also, 'in the UK researchers have a duty of care under the 1989 Children Act to report instances where they have reason to believe that a young person is in danger from others or is likely to cause danger *to* others' (Heath *et al.* 2009: 35). Research may reveal that a young person is being abused and so in such a situation, a researcher is duty bound to pass on such information to a responsible adult. Clearly, such issues are very sensitive and must be handled with much care.

The negotiation of confidentiality varies according to the focus of research and the needs of the young people involved. In my own work with unaccompanied minors, I found that many of the young people regularly raised different concerns and issues with me. Although I always encouraged them to think through these issues to reach a conclusion, there were also cases where I felt compelled to speak to specific service providers about their concerns. I did this only with the permission of the young person concerned and clarified with them that

Box 3.3

Breaching confidentiality

Reasons why you may decide to pass on information about a young person:

- You think that passing on information could help to solve a young person's problem.
- You want to allay your anxieties about a young person's well-being by sharing them with someone else.
- You see it as your professional duty to act in compliance with child protection laws and good practice guidelines.
- You have been told something that you would want to know if you were the young person's parent or teacher.
- You do not want to be held responsible later if something goes wrong.
- You want to maintain good working relationships with other professionals (for example, teachers, social workers) and see information-sharing as vital to this.
- Gatekeepers have requested that you pass on certain kinds of information to them.

Options which may not involve breaching agreements about confidentiality:

- Initial discussion of concerns with a colleague in which no names are mentioned.
- Further discussion with line managers or supervisors.
- Calling a confidential telephone helpline to discuss your concerns and possible options.
- Writing a full account of the incident, to be kept on file or passed on to social work, police or a child protection officer if working within an organisation.

Adapted from Gallagher (2009a: 21–2)

I would only mention the specific issue raised and that the remainder of the interview would remain confidential. I concluded that

> it is clear then that researchers should be aware of the principles of public and social network confidentiality, the boundaries of limits of each, and be willing to discuss these matters clearly and openly with the children they are doing research with in order to ensure that any harm or risk is minimised.
>
> (Hopkins 2008b: 43)

Incentivising participation

An additional set of ethical and methodological considerations in research with young people relates to the increasingly popular practice of offering incentives to young people who participate. As Heath *et al.* (2009: 36) note, 'it is becoming increasingly common in youth research to offer some form of material benefit to young people in return for their participation in research.' With regard to her work with working-class young men in the UK, Linda McDowell (2001: 90–1) notes:

> I explained that it would involve meeting me three times over the next 12–18 months and also that I would pay them a small sum at each meeting. I strongly believe that it is important to be able to recompense the individuals who are prepared to answer what must often seem like intrusive questions from social scientists. As these young men were all from relatively low income households and many of them held casual jobs whilst still at school, a small sum for participation seemed both an adequate reward and hopefully a way of encouraging their participation over the course of the year or so. Other researchers engaged in youth work have 'paid' their participants in kind, using vouchers for popular stores for example.

Although McDowell (2001) paid cash to the young men, other incentives commonly used in research include vouchers for music, books or high street stores, entry into a prize draw or credit towards university courses (Heath *et al.* 2009). However, there has been a lack of consideration given to the complex ethical issues that arise given the adoption of such practices by researchers. Malcolm Hill offers a balanced perspective when he notes that views differ with regard to the appropriateness of giving money or gifts to young people as a result of their participation in research. 'Some view this negatively as inducement or bribery. Alternatively it can be seen as fair recompense' (Hill 2006: 71). Moreover, as Hill (2006) points out, if adults are given payment in a study, then giving payment to young people can demonstrate that each group is equally valued. Furthermore, it may now be the case that young people expect to be paid for participating in research which has worrying consequences for researchers working within the constraints of limited budgets yet conducting valuable and important research.

A clear distinction is often made between offering payment upfront to young people participating in research compared with young people receiving a reward after they have taken part (Heath *et al.* 2009, Hill 2006). Knowing about incentives before participating in research may encourage participants to take part when they do not fit within the remits of a project. For others, it may encourage them to say what they think the researcher wants them to say for fear of having their reward withdrawn. Likewise, researchers who choose to reward young people after they have participated in research may find additional volunteers once they have heard that their peers were rewarded for taking part (Heath *et al.* 2009). The key issue here is to think carefully about the possible knock-on effects of rewarding young people for participating in research rather than just assuming that paying young people or giving them vouchers is an ethically sound practice.

Positionalities and power relations

Issues related to power relations are often flagged up as an important consideration in youth research as a result of the marginal position that many young people hold in society. Connected with sensitivities to power relations, the multiple positionalities of the researcher are also important to reflect upon in the conduct of ethical research. These multiple relations of power are important to consider not only in relation to the young people participating, but also in connection with gatekeepers and other adults who may be involved in the facilitation of access or who may be present during particular aspects of the research.

Recognition of the multiple power relations (Matthews 2001) existing within the research context helps to improve the validity and reliability of research as the researcher is more likely to be aware of the different structures of power that may be influencing young people's participation. Often – although not always – the researcher is in a position of power compared with the young people who are taking part in research. This is not necessarily a problem and rather is something to be aware of in research encounters. There may be ways in which the researcher could attempt to minimise their power. For example, young people could be involved in the design and implementation of the research rather than just the researcher alone. The researcher could choose to sit on the floor or use body language and behaviours which are less threatening to young people.

The power imbalance that often exists between adults and young people can be further complicated by the intersecting positionalities of the researcher and the young people. As such, young people's identities and those of the researcher often operate in ways that create new forms of power relations, senses of difference or feelings of connection. Young people may possess particularly marginalised forms of identity associated with their social class, education, family background or ethnicity. Alternatively, they may feel empowered by their membership of particular dominant social groups associated with their gender, class, schooling or their participation in particular leisure or sporting activities. Similarly, although the researcher will normally be an adult, they will also be gendered, classed, aged and so on. This means that although there may be a difference on the basis of age, there are also likely to be other differences and similarities according to the other identities of the researcher and the young people being consulted. Examples of researchers considering their positionality in research include John Horton's (2001) discussion about some of the anxieties associated with his male identity and his status as a youth researcher and Tracey Skelton's (2001) reflection about her class background in the context of doing research with young women (for further discussion about positionality in research with young people, see Hopkins 2007d, McDowell 2001).

Key themes

- Ethical issues should be central to the conduct of good research about young people, place and identity. It is important to be respectful and fair, to minimise harm and to promote the social well-being of the young people involved.
- Conducting ethical research with young people requires careful consideration of a range of contextual and methodological factors at all points of the research process and not only at the start.
- Although there are a number of ways in which researchers can conduct ethical research about young people, place and identity, it is also important to be aware of practices that are regarded as unethical.
- Informed consent is important in ethical research about young people, place and identity, and relies upon young people being as fully informed as possible about the details of the research project.
- Confidentiality and anonymity are also important in ethical research about young people, place and identity.
- Although it is often deemed ethical and appropriate to incentivise young people's participation in research, it is also important to be

sensitive to the complexities and exclusions that such practices can create.

- The multiple positionalities of the researcher and researched raise additional ethical issues which it is important to be sensitive to.

Project ideas

Find a research report and journal article about young people and examine the approach the researcher adopted in seeking to obtain informed consent from the young people involved in the research.

Discuss critically the ethical issues that arise from incentivising young people's participation in research.

Discuss critically the main ethical considerations involved in research about young people, place and identity.

Suggested further reading

This excellent book includes details of numerous ethical issues that arise in research with young people, including guidance about how these issues can be managed:

Alderson, Priscilla and Morrow, Virginia (2004) *Ethics, social research and consulting with children and young people*. Barnardos: Barkingside.

This is a very useful set of articles about ethical issues in research with children and young people in a diverse range of contexts:

Children's Geographies 6(1) Interdisciplinary perspectives: ethical issues and child research.

This is an interesting set of short commentaries about a range of ethical issues that researchers encounter in work with children and young people:

Ethics, Place and Environment 4(2) 117–18, series of short communications about the ethics of working with children and young people.

Chapter 2 of this book explores a range of ethical issues in research with children and young people and includes useful case studies:

Tisdall, Kay, Davis, John and Gallagher, Michael (2009) *Researching with children and young people: research design, methods and analysis*. London: Sage.

Part II

Scales

4 The body

Emilia: . . . always change my mind about how I want to look depending on what I'm wearing and that sort of thing. I wish I had three different bodies I could change into. Sometimes it gets you down. You get depressed about yourself in general it's just something you have to live with and everyone's in the same boat. It's not that much of a major problem. (Budgeon 2003: 35)

Sameera: I'm constantly thinking about what people will think of me, they must think that I'm really typical . . . even when I haven't got a scarf on my head but I'm like in Asian clothes I'm so *paranoid*. Oh people must think typical . . . you know that I'm from the dark ages and that . . . I know it sounds bad but that's how I feel. (Dwyer 1999a: 12)

Michelle: It's my confidence. I struggle to talk to new people. Most people take one look at you, and they jus divvn't wanna know. I canna be bothered with that. I'd rather just stay round here – everyone knows me, I get nee bother. (Alexander 2010: 73).

Youthful bodies

Emilia, Sameera and Michelle are all talking about how they experience, manage and cope with their bodies, the ways their bodies are presented and how other people respond to their bodies. Emilia reflects upon how her emotions are intertwined with what she chooses to wear and Sameera considers the possibilities of what stereotypes people will reinforce about her as she chooses to wear a headscarf and dress in a manner that reflects her parents' cultural heritage in South Asia. Michelle's perspective on her body is connected to her desire to manage her gendered identity and working-class background in the context of the different ways in which she is treated in her local neighbourhood and city. Clearly, for these three young people – as for many young people, their bodies are highly charged sites of identity construction.

The body represents one of the most significant locations when thinking about the intersection of young people, place and identity. In the most basic sense, it is through young people's bodies that they identify themselves, and/or are identified by others, as being young people. In the introductory chapter, processes of identification were discussed (see Box 1.1) highlighting five mechanisms whereby identities can be understood, and all of these can be applied to understanding young people's bodies and their interactions with different spaces and social identities. First, identities may be actively conveyed of communicated, so a young person may actively communicate their age through their body. Second, identities can be read or re-interpreted by others, and so young bodies can be read by others and interpreted in a variety of ways. Third, identities may be negotiated – sometimes adopted, modified, resisted or contested – in different ways and so young people may negotiate their embodied identities in different ways according to where they are and how they feel. They may challenge interpretations, resist belonging to particular social groups or modify their bodies in specific ways in the articulation of their identities. Fourth – and of particular relevance to this chapter – identities may be performed through particular practices such as body language, dress or other bodily markings, and so young people may wear particular clothing in order to express who they are and where they belong. Finally, identity processes can work to take over, occupy or reshape places and so young people's articulations of their identities in particular places can gain a sense of occupancy with places and sometimes lead to places being made and remade in diverse ways.

Young people's bodies are the locations where they express their identities through clothing choice, hairstyle and perhaps other bodily markings, such as jewellery, make-up, tattoos and other accessories. Simultaneously, young people's bodies are the main sites where their identities are interpreted: where they are read and understood as 'young people'; how they are assumed to act and behave in particular ways and possess specific qualities; and where they might be regarded by others as appropriately occupying particular spaces, yet may be restricted, regulated or even prevented from entering, other places. 'Seen as both *risky* and *at risk*, young people's bodies become markers of the state of the social body now and in the future. They become emblematic of a series of societal fears and concerns, calling up deep-seated fears about the loss of romantic childhood' (Hörschlemann and Colls 2010: 4, see also Colls and Hörschelmann 2009). Young people's bodies can be marked out, read, interpreted and responded to in a wide variety of ways, and given this complexity, I only hope to offer a brief insight into research about young people's bodies in this chapter. Initially, I provide some background to understandings of what is meant by 'the body'. Then, I focus upon three main areas of work in order to highlight the significance of the body in thinking about young people, place and identity including: *marginalised* bodies, *subcultural* bodies, and *excessive* bodies.

Despite the interest in the body, bodies and embodiment across the social sciences, very few researchers have offered a concrete definition of what 'the body' is. In thinking about 'the body', Ruth Butler (1999: 239) notes that 'the human body is not just flesh and blood. An object for the mind to use at its will. The body is an active and reactive entity which is not just part of us, but is who we are'. Robyn Longhurst (2001:11) also notes that, 'we all have one, or at least, we all are one. We are all born, we all die. Although these things appear to be universal our embodied experiences are unique.' This uniqueness is due to the different ways that bodies are differentiated according to the social categories, assumptions and values that societies impose on different bodies. Ruth Butler (1999: 238) emphasises that 'society's narrow expectations of any particular body are usually closely allied to the crude and broad social classifications of gender, age, race, ability, class and sexuality into which they are placed.' Likewise, Robyn Longhurst points out that 'everyone has a body (indeed, *is* a body) but bodies are differentiated through age, ethnicity, sex, sexuality, gender, size, health, and so on'. Similarly, Beverley Skeggs (1997: 82) notes that bodies are 'the physical sites where the relations of class, gender, race, sexuality and age come together and are embodied and practiced', and Stuart Aitken (2001a: 62) explains 'bodies are inscribed by different race and ethnic backgrounds, by their age, sexuality and able-bodiedness, and by where they reside, work and travel'. So, bodies are sites in themselves, but are also locations where social identities are marked out and practised, and so bodies are key locations in understanding the complexities of young people, place and identity.

Elizabeth Kenworthy Teather (1999: 7) notes that 'bodies occupy space, but they are also spaces in their own right' and that bodies are encoded with different emotions and so can be 'maps of desire, disgust, pleasure, pain, loathing, love.' The ways in which bodies are read, interpreted and responded to are therefore related to the ways in which different bodies are associated with power, authority and control or with weakness, marginality and submission; there are emotional geographies. So, although bodies regularly associated with a number of social identities, the meanings and stereotypes associated with these identities result in bodies being associated with, and responded to, according to certain emotional and expressive responses. Again, Ruth Butler (1999: 241) notes that 'understandings of and reactions to people are culturally created and perpetuated. The ways that we react to people, what we expect or do not expect of them, where we expect to see them, what we expect to see them doing are all issues affected by images of people' which are entwined with everyday social and cultural relations.

Overall, Gill Valentine (2001: 15) summarises this discussion, by offering a definition of what the body is:

> It marks the boundaries between self and other, both in a literal physiological sense but also in a social sense. It is a personal space. A sensuous organ, the site of pleasure and pain around which social definition of wellbeing, illness, happiness and health are constructed, it is our means for connecting with, and experiencing other spaces. It is the primary location where our personal identities are constituted and social knowledges and meaning inscribed. For example, social identities and differences are constructed around bodily differences such as gender, race, age and ability (Smith 1993). These can form the basis of exclusion and oppression (Young 1990a). The body then is a site of struggle and contestation. Access to our bodies, control over what can be done to them, how they move, and where they can or cannot go, are the source of regulation and dispute between household members at work, within communities, at the level of the state, and even the globe.

So, the body then is who we are. It is part of our sense of our everyday emotions, and is how we connect with and experience the world. It is where our identities are, and where the negative and positive assumptions about these are placed. Our bodies are also regulated in particular ways and can become sites of dispute across a variety of scales. In the context of the explosion of interest in the body, Stuart Aitken (2001a: 62) notes that 'young people's labor and their bodies are also largely missing from the debate'. This chapter explores the contested bodies of young people, and in doing so attempts to highlight the social, cultural and political issues tied up with processes of embodiment experienced and articulated by young people. After all,

> [m]aking bodies explicit in our theorizing and empirical investigations has provided new ways of understanding subjectivities, power and politics. It has allowed us to develop a deeper understanding of the ways in which individuals, communities, and nations are simultaneously engaged in various mutually constituted power relations in different places and at different times.
>
> (Longhurst 2005: 345)

Marginalised bodies

There is a range of complex ways in which the intersection of particular social identities in different places shape – and are shaped by – young people's bodily markings, gestures, clothing and other embodied practices. Recent research about the experiences of ethnic minority young people highlights the significance of the body as an important site for the ways in which their race, ethnicity and religion are expressed, judged and responded to by society. Some of the most interesting work in this area has focused on the embodied experiences of Muslim young men and women. Here are three statements from young Muslim men involved in research that I conducted in Scotland (Hopkins 2004: 262):

Rehman: I mean . . . to walk around the streets in like traditional Muslim dress, I think people would have looked at you and wondered what you were up to and would be suspicious . . .

Asadullah: Nowadays when people see the word Islam, like as soon as they hear about it, or see someone with a beard, or whatever, they just think of terrorists. They don't think of what Islam should be about, they just think of terrorists, that's what they do.

Aslam: I reckon racism in general has changed since then . . . if you're coloured, you know if you're a Muslim, you're related to Al Qaeda.

It was clear from the focus group and interview discussions with these young men that a series of events, and in particular the terrorist attacks in New York on 11 September 2001, had resulted in their bodies becoming intensified sites for how their identities are read and responded to by society. As I noted:

> According to the young men, markers of 'Muslimness' have heightened in significance as a result of the events and aftermath of September 11th 2001. In particular, dress choice, young men with beards and skin colour are recognised by the young men as markers that identify people as Muslim, and therefore as a threat. These markers have a powerful influence in determining the experience of young Muslim men's everyday lives in Scotland.
>
> (Hopkins 2004: 261)

Clearly then, the marginalised bodies of young Muslim men are being increasingly stigmatised according to the ways in which bodily markings and other features – such as skin colour, dress choice and facial hair – combine to identify them with a particular religious group and work to stereotypically associate them with particular threatening practices. This indicates that religious identities are being racialised, as assumptions based on race (i.e. making biological assumptions about individuals based on phenotypical features such as skin colour) and ethnicity are being interpreted as being connected with, or the same as, particular religious identities. This racialisation of religion has also been highlighted by research conducted in Australia (Dunn *et al.* 2007). Likewise, in their book *Kebabs, Kids, Cops and Crime* (2002), Jock Collins, Greg Noble, Scott Poynting and Paul Tabar include discussion about 'Middle Eastern appearances' pointing to the ways in which young people's bodies are read and responded to according to assumptions about race, national origin, religion and location.

What has been discussed so far focuses on the embodied experiences of young Muslim men. Clearly young Muslim women's bodies are also contested sites in terms of the construction and contestation of identities (Dywer 1998). This is demonstrated in Box 4.1 in which the embodied experiences of Muslim women in the UK and USA are discussed, revealing that their bodies are possibly more intensified sites of identity construction and contestation than for young Muslim men.

A second example of young people's marginalised bodies can be found in research about disabled young people. Although many young people have hidden disabilities, others have disabilities which are not hidden, and may influence the bodily presentation of an individual. As a result

Box 4.1

Contested bodies of young Muslim women

Research in both the UK and USA has highlighted the complex ways that young Muslim women's bodies are contested sites for the construction of their religious, ethnic and gendered identities. This work focuses upon the ways in which skin colour, gendered identities and – in particular – dress choice intersect to mediate the experiences of young Muslim women. This research explored how young Muslim women managed the stereotypical divisions between being Asian or being English, and the embodied practices and behaviours associated with this. Humaira, one of the research participants, tends to wear 'English clothes' and this leads her classmates to question her religious identity:

> They all say 'Oh you can't be one of us because you look totally different' . . . One girl was asking me 'Are you a Muslim?' And I go 'No I'm from Venus!' And she asked her friend and her friend told her where I was from. And she goes 'Oh it's just that you don't look it.'
>
> (Dwyer 1999a: 10)

This work has also shown that for 'young British Muslim women dress is a powerful and overdetermined marker of difference' (p. 5). Through exploring 'the rhetoric of the veil', Dywer (p. 18) found that the young women argued that it meant 'nothing about the sexual and moral propriety of the wearer', as young women would wear the headscarf strategically to navigate different contexts – such as wearing it on the way to and from school but removing it whilst at school. Alongside these negotiations associated with dress choice, Dwyer (pp. 20–1) also argues that the wearing and not wearing of the headscarf was 'part of the process of experimenting with alternative (sexed and gendered) subjectivities'.

More recently, research in the USA by Mayida Zaal, Tahani Salah and Michelle Fine (2007) has explored the ways in which young Muslim-American women negotiate their embodied identities. Like Dwyer, they found dress to be an important issue for the ways in which particular bodies are marked out and assumed to have certain qualities. They comment that 'many of the young women worried for mothers, sisters, brothers, uncles who are easily identifiable as Muslim-Americans because of their facial features or style of dress and for those who try be less visible by changing their ethnic Muslim-sounding names to avoid the humiliation and stigma'.

Adapted from Dwyer (1999a) and Zaal *et al.* (2007)

of poor media coverage that either ignores or shames disabled youth and the often negative attitudes and behaviours of others, disabled young people can feel embarrassed and ashamed by their stigmatised bodies. Furthermore, the placement of some disabled youth in 'special schools' also works to further marginalise them as their bodies become excluded from 'mainstream' education. Sally French and John Swain (2004: 199) note:

> Young disabled people grow up in a disabling environment where they face numerous physical and social barriers on a daily basis. These may include inability to take part in leisure pursuits, hostile attitudes or avoidance from other children, patronising behaviour from adults and exclusion from mainstream school. The formative people in their early lives are not usually disabled themselves and may unwittingly pass on stereotypes and ideas about disability which affect adversely the young person's self-image and self-confidence. Such ideas may include the inability of disabled people to work, to become parents, or to have satisfying sexual relationships.

Furthermore, the particularities of a young person's disability may result in them possessing particular bodily characteristics or movements which they have limited control over, and so the actions they may want to take could be impossible to carry out. Disabled young people may also experience bodily pain or discomfort associated with their disability and such everyday challenges are often heightened by the negative ways in which disabled youth experience social relations. Ruth Butler (1999: 85) observes that 'disabled youths find themselves objectified by the curious gaze of strangers' and Zoebia Islam (2008: 46) discusses 'two young people's experiences of being gazed at . . . and the feelings of embarrassment and invalidation and anger and frustration this led to'. Islam (2008: 46) refers to the experience of Aisha – a young disabled women – who feels that her body is marginalised to the extent that it is ignored by others:

> When I'm shopping with mum, old people . . . old Asian people come over and say have you got eczema. Like they'll just come and ask my mum like I'm not really there and I'm like 'I'm here Hello! I'm standing right next to you'. They just think we, disabled people, are not there, we don't exist.

Islam (2008) also found that young people also identified with various other identities and some actively resisted being labelled 'disabled' and instead linked with ethnic, religious and gendered identities or countries associated with their cultural heritage.

Another group of young people who arguably possess marginalised bodies are chavs. The term 'chav' (or 'ned' as used in the west of Scotland in particular) is sometimes associated with the Romany word for child (chavo or chavi) although it has also been suggested that it emerged from the north-east of England word 'charver' (Nayak 2006, Tyler 2008). As Imogen Tyler (2008: 21) confirms, the use and associations with the term have grown considerably over the last few years:

> Chav, and its various synonyms and regional variations (including Pikey, Townie, Charver, Chavvette, Chavster, Dumbo, Gazza, Hood Rat, Kev, Knacker, Ned, Ratboy, Scally, Scumbag, Shazza, Skanger), have become ubiquitous terms of abuse for the white poor within contemporary British culture. Since 2003, we have seen the emergence of an entire slang vocabulary around chav, which includes terms such as chavellers cheques (giro and benefit payments), chavtastic, chaving a laugh (laughter at chavs), chavbaiting, chavalanche (large group of chavs), chavallier (chav car), chavspeak, chavspotting and acronyms such as 'Council House and Violent', 'Council House and Vile', and 'Council House Associated Vermin'.

This particular group of young people is often identified by their embodied practices such as their dress, accent and general conduct. As Anoop Nayak (2006: 822) confirms, 'by adopting the outward manifestations of street style – baseball caps, tracksuits, trainers, heavy gold jewellery – and accompanying this apparel with a pronounced walk, Charver lads were engaged in a body-reflective technique of "hard" masculinity.' For this reason, some would argue that chavs are a youth subculture (see Table 4.1). However, as it is unlikely that chavs would self-identify as such, it could be argued that they are not necessarily a youth subcultural group in themselves.

Young people who are or are identified as being a chav often experience very explicit forms of discrimination, so being part of this group can further the marginalisation they already experience as clothing is used as a signifier for marking out a lower-working-class background and its associations with unemployment and school exclusion (Alexander 2008, 2010). Place is often important to the formation of identities within such groups as confirmed here:

> The Chavs/Neds showed important features of a subculture as they all expressed a particular style of dress, and for many outside the

> group their behaviour was considered to be the same, often viewed as deviant. Through dialogue with members of this 'perceived' subculture, it became clear that this group were in fact not homogenous, and that their affiliation with a group was determined by their locality.
>
> (McCulloch *et al.* 2006: 554)

Linked with the important of place, Imogen Tyler (2008: 17) argues that chav is an explicitly classed term and is part of a broader insecurity about social relations. Likewise, Anoop Nayak (2006) argues that chavs are racialised, classed and criminalised in ways that reflect social anxieties around issues such as fear of crime. As such, chav 'has become a ubiquitous term of abuse for white working-class subjects (Tyler 2008: 17). Furthermore, the term is gendered, as shown by research conducted by Imogen Tyler (2008) in which she explored the representation of chav culture through the character-led scenes of Little Britain, a very successful BBC comedy series (2003–6) (see also Alexander 2010). Characters such as Vicky Pollard contribute to the association of terms like 'Chav mum' being produced through 'disgust reactions' which are about an 'intensely affective figure that embodies historically familiar and contemporary anxieties about sexuality, reproduction and fertility and "racial mixing"' (Tyler 2008: 18). At different times and places, these images 'become over determined and are publicly imagined (as figured) in excessive, distorted, and caricatured ways' (Tyler 2008: 18), and so the bodies of chavs move through the multiple places of society and social relations provoking a series of repeated disgust reactions (Tyler 2008).

Not only can chavs be identified by their clothing and style, but dressing like a chav, or wearing particular clothes identified with this group can restrict where a young person might gain access to. Anoop Nayak found that chavs had 'been priced out of many of the new drinking venues' in the post-industrial city and were further marginalised by 'their particular style of clothing, which included tracksuits, trainers and baseball caps', all of which were banned from these places, so they were therefore forced to find their own places within the city which normally involved hanging around street corners 'drinking cans of beer, smoking and chatting to friends' (Nayak 2006: 820). These young people are therefore spatially segregated from other young people as well as being distinguishable on the basis of both their dress and demeanour. There have been some attempts too, to reclaim 'the term as an affirmative sub-cultural identity' (Tyler

2008: 31) with usage of the phrase being associated with a form of social discrimination and exclusion. The significance of the intersection of youth, place and identity is summed up here by Imogen Tyler (2008: 32):

> The figure of the chav is mobilised in ways that justify the continued division of society into those who can speak, act, and feel and those who are 'spoken for'. Being identified as 'chav' not only means having the place you live, the way you speak, the clothes you wear, your culture, habits, and lifestyle subject to perverse misrepresentation, mockery and derision but also actively blocks your social mobility.

Subcultural bodies

From 1964 until 2002, staff at the Centre for Contemporary Cultural Studies (CCCS), based at the University of Birmingham, conducted research and produced numerous publications focused on youth subcultures (see, for example, Hall and Jefferson 1976, Hebdige 1987, Willis 1977). The subcultural approach to understanding young people's lives is largely associated with the work developed in this Centre, which the university controversially closed down in 2002. In terms of defining 'youth subcultures', Mary Jane Kehily (2007b: 21) confirms that

> the concept of 'subculture' is important in understanding the social lives of young people. In simple terms a subculture can be seen as a group within a group. The social group frequently referred to as 'youth' has thrown up many subgroups over the years which come to be regarded as subcultures. Over time, these subcultures acquire names and identities such as teddy boys, skinheads, punks and goths.

At different times and in different places, specific youth subcultures have become the subject of negative media coverage, suspicion by the public and often associated with deviant forms of behaviour. New subcultures often become labelled as new 'folk devils' as moral panic ensues with regard to their visibility, behaviours and intentions (Widdicombe and Wooffitt 1995). Table 4.1 provides some examples of different youth subcultures.

Youth subcultures are inclined to see themselves as different from other

Table 4.1 ***Some examples of youth subcultures***

Biker	Emo	Gamer	Gangsta	Goth
Hippie	Jock	Mod	Punk	Preppy
Raver	Redneck	Rocker	Skateboarder	
Skinhead	Straight Edge	Teddy Boy		

cultures, including those of other youth cultures and the cultures they experience at home (Kehily 2007). Identifying with a particular subculture often relies on similarities being developed between young people in terms of their financial situations, responses to similar issues, as well as the likelihood of them having more leisure time than adults and being the consumers of pop music (Wyn and White 1997). The majority of youth subcultures identify with particular ways of dressing, styling and managing young bodies, and so as these subcultures become more widely recognised in society, young people are often identified by others as belonging to particular youth subcultural groups. As Wyn and White (1997: 73) observe, 'cultural formation is an active process which involves the participation of young people, and which marks out different relationships to the dominant ideologies and values of society'. It is also important to note that although some young people may identify with a particular subcultural group, 'sole identification with one particular group or style in a manner which is rigidly fixed over time is relatively unusual' (Wyn and White 1997: 83). This suggests that young people may identify with a particular subcultural group yet may change affiliations over time or may simultaneously belong to two or more groups. Other research has suggested that young people are very aware of the boundaries between different subcultural groups, that there is no joint membership and that it could not be chosen freely (McCulloch *et al.* 2006), thereby suggesting that boundaries are more rigid. So, although the concept of youth subcultures offers a useful insight into the identities of young people, it is not as useful at explaining why and how different groups behave in particular ways, or how and why they act differently from other groups (Wyn and White 1997).

According to Brian Bailey (2006: 338), Emo – which is short for 'emotional music' – is 'an evolving and complex American youth subculture that listens to a specific genre of music, which is characterised by feelings of vulnerability and a willingness to express heartfelt confessions about adolescence.' It emerged from the music scene in Washington DC, USA in the 1980s, although arguably today there are young people who identify as – or are identified as being – Emo young people in a diverse range of places including Australia, New Zealand, Singapore, Japan, Mexico, Canada and various countries in western Europe. Like many youth subcultures, the exact definition and boundaries of what constitutes Emo subculture is contested, changes over time and – given its geographic spread – varies across places as well. Brian Bailey (2006: 338) comments on 'the various and sometimes conflicting social practices associated with Emo subculture' and observes how it has become a 'highly contested set of meanings and collective practices'. Bailey (2006) suggests that there have been splits within Emo subculture with original members of it forming new subcultures such as 'Emo independents', 'Emo mainstream' and 'Emo kids'. This diversity is further highlighted by the fact that

> [t]o some, it appears to be a cathartic experience through a genuinely outward release of painful emotions coupled with a sense of grace, self-pity, and hope. For others, it means rejecting the music industry hegemony for a DIY (do-it-yourself) lifestyle and following a band that seems like 'your own little secret.' For many kids, it means behaving in a way that respects people's feelings, and to others, it means striving to look like their favourite emo band's lead singer whilst singing along at a concert.
>
> (Bailey 2006: 338)

Like many youth subcultures, Emo young people are often associated with certain fashions, although it is debatable to what extent the fashion or the subculture influenced each other or emerged independently (Wyn and White 1997). However, it is clear that certain embodied practices and ways of clothing and managing the body are associated with Emo young people. Stereotypically, an Emo young person might be expected to wear tight jeans and t-shirt, studded belt, converse shoes and black thick-rimmed glasses and have pierced lips, eye liner and long straight black hair with a fringe brushed to one side. Stereotypical behaviour of Emo young people associates them with being shy, introverted, quiet and perhaps also nervous, somewhat depressed and insecure. For Emo young men in particular, this can lead to them receiving homophobic remarks due to their clothing choice and style.

Excessive bodies

Just as young people's bodies are often seen to be marginalised or subcultural, so too are they often seen to be excessive. Whether it be related to their behaviours, attitudes or values, or their style, fashion or general demeanour, young people are often stereotyped as being excessive, over the top, out of the ordinary, unwarranted, strange and a challenge to the status quo. Research about young people, place and identity has explored numerous ways in which young people's bodies might be regarded as excessive, but here I focus upon two specific examples. With the first example I focus upon the bodies of young people who are often stereotyped as dancing and drinking to excess. Here, I explore the different ways in which young clubbers' bodies are prepared, judged and experienced during the course of a night out. Second, I draw attention to young people who are seen to have excessive body weight or fat by highlighting the experiences of obese young people.

Ben Malbon (1999: 16) conducted an excellent ethnographic study about the experiences of going clubbing, and – connecting with the discussion about culture above – he confirms that 'the latest and by a long way the largest and most influential of recent young people's cultures or styles in Britain can be found in club cultures'. His exploration into 'the night out' regularly refers to the salience of people's bodies in such experiences, and the highly embodied nature of going clubbing. One of the key moments and heightened situations of a night out can be negotiating entry to a club by managing to convince those on the door that you should be admitted to the club. For some, this might be an aspect of a night out that passes without much notice, yet for others, this can be the moment that ends their night out as their friends gain entry and they do not. Gaining entry to a club is a highly embodied process, as a series of judgements is made, very quickly, by those on the door, about whether or not entry should be granted. For clubbers, then, this requires presenting their body in a particular way. As Malbon (1999: 64) confirms, 'negotiating entry can certainly demand the display of a "correct" style, the deployment of a very specific set of practices and a successfully employed "personal front".' However, it is not only about presenting a 'personal front' but possessing particular social identities that can result in securing entry, as it is clear that certain social identities are regarded as more desirable than others:

> In addition to enforcing judgements about coolness, bouncers on 'the door' at clubs can also reinforce wider societal prejudices in respect of ethnicity, gender, and 'good looks' – some clubbers find entry more problematic than others.
>
> (Malbon 1999: 64)

This research found that black men often find themselves barred from clubs or subject to quotas where a limited number of black men are allowed into particular clubs at any one time. Furthermore, other clubbers claimed that various forms of discrimination take place, even although policies varied from night to night. Malbon (1999: 67) refers to negotiating the door as 'a very explicit form of "identity-exposure" ' and compares it to an 'attempt to pass a test'. Locating the negotiation of the door in a series of other decisions that take place during a night out, he notes:

> Yet the trials of the door appear more than worth paying. A clubber queuing to enter a club has already decided that they are 'right' for the club and think that they *could* belong. The door becomes the moment when certain facets of the identity that a clubber has constructed for himself or herself are used to gain access to a social situation in which other facets of that clubber's identity may be – relatively speaking – submerged or unschackled. Thus, these early and pre-clubbing stages of the night are significant because they provide a test of identity, and the passing of this test – the validation of the clubber's identity on the way through the door – can act to establish and reinforce the emergent belongings through which the clubbing crowd inside is bound together.
>
> (Malbon 1999: 68)

So, in the same way that young people's marginalised and subcultural bodies are managed, styled, judged and accorded restricted access to particular places, the same can be said for those who go clubbing. However, the importance of the space of the body does not necessarily diminish after gaining entry to a club. Negotiating bodies through clubbing crowds, dancing in the dark and other emotional and embodied experiences associated with dancing, drinking, music and drugs can all form part of the clubbing experience, as highlighted in 'the embodied practice of "coolness" ' (Malbon 1999: 57), that clubbers might engage with in an attempt to attract attention to themselves.

Alongside this, clubbers may experience particular emotional geographies such as those associated with the 'oceanic' or the 'ecstatic' (Malbon 1999). The 'oceanic' refers to 'those experiences characterised by one or more of these sensations: ecstasy, joy, euphoria, ephemerality, empathy, alterity (a sense of being beyond the everyday), release, the loss and subsequent gaining of control, and notions of escape' (Malbon: 1999: 107) which can be either drug or non-drug induced emotional and embodied sensations. On top of this, the 'ecstatic' is 'an additional layer of emotional and sensational action' (Malbon 1999: 116), brought about through the use of drugs. Again, these embodied sensations are not unitary and instead vary during a night out as various stages of sorting and preparing the drugs occur before they are taken, after which there may be experiences of losing touch, elation, euphoria, intensity and then a sense of withdrawal. So, during a night out clubbing, the body is a key site for the negotiation of various identities and practices, and these embodied emotions and experiences change, alter and are re-shaped during the course of the evening, and often rely on the bodily practices of an individual young person, alongside judgements made by friends, bouncers and the clubbing crowd.

Young people's clubbing behaviours are often stereotyped by society as being excessive and over-the-top, with media coverage regularly linking clubbing (and related activities) to problems with drinking and drunkenness in public places. Clubbing therefore often works to further stigmatise and alienate young people from society. However, some of the research participants in Malbon's (1999) study argued that their participation in clubbing was a key method whereby they temporarily avoided the stresses and strains of everyday life, and some even saw their clubbing as a form of resistance to mainstream society. Just as Mark Jayne, Sarah Holloway and Gill Valentine (2006: 464) have suggested about drinking and drunkenness, it could be that we should consider clubbing as giving young people an opportunity for them to create their own places within the city, and 'assert their identities as active participants in consumer society'.

A second category of young people who could be regarded as having excessive bodies is young people who are regarded as fat, overweight or obese. Indeed, it is widely recognised that in many Western societies, people are generally becoming more overweight. Given that young people are often regarded as adults-in-the-making, this specific group of young people is regarded as particularly problematic due to medical concerns that suggest that obese young people are more likely to become obese adults, and therefore suffer from a range of health problems as a result of being overweight:

> Adolescent health has implications for the health of future populations. Obese adolescents are likely to remain obese throughout their adult lives, have poor health and reduced life expectancy through increased risk of associated diseases.
>
> (Lake and Townshend 2006: 265)

Although the supposed 'escalating obesity epidemic' (Lake and Townshend 2006: 262) may be a key concern for the medical and health professions, very strong medicalised discourses that promote negative representations of overweight young people have negative outcomes for many young people as a result of the ways in which their bodies become the source of ridicule, embarrassment and humiliation. Susan Smith (1993: 62) has referred to 'the medicalisation of social life' which she defines as the

> drawing of boundaries around social groups on the basis of presumed health, illness and susceptibility to disease. Such boundaries are overlaid with attitudinal, behavioural and territorial markers, and may be used as criteria in determining the differential apportionment of goods and services.

As such, young people who are overweight are often stereotypically associated with particular behaviours and attitudes as a result of assumptions about their health and well-being, and this can influence where they can and cannot go. David Bell and Gill Valentine (1997: 29) have noted that there is a dominant discourse 'where a fat body is understood to be unhealthy, ugly and sexually unattractive', and so 'fat people are stereotyped as undisciplined, self-indulgent, unhealthy, lazy, untrustworthy, unwilling and non-conforming' (1997: 36). Furthermore, these issues are tied up with the ways in which governments promote healthy eating as part of citizenship, and so overweight young people are stigmatised as being less worthy citizens compared with other young people (Rawlins 2008). This then results in particular bodies being stereotyped in different ways, often along the lines of social class: 'tasteful and tastless bodies mark class dispositions in particular ways, of course, as the disdain for obesity in postmodern Western societies vividly illustrates' (Holliday and Hassard 2001: 3). These stereotypes of distaste and disgust are intensified by 'media images, and particularly fashion photography' which has resulted in the perpetuation of 'mythical stereotypes of the perfect female form as young, beautiful and thin' (Crewe 2001: 633).

Many of these stereotypes can also be reinforced in young people's lives through particular institutional spaces and government policies and priorities. Recent research in the UK has found that policies such as the Healthy Schools Programme and the related focus on exercise and healthy eating can heighten the marginalisation experienced by overweight young people, making them likely to be the targets of bullying (Curtis 2008: 414):

> Secondary schools can therefore be particularly challenging environments for young people with obesity. School-based PE and the promotion of 'healthy eating' can accentuate the otherness of the overweight child and create opportunities for surveillance and persecutory behaviours from peers.

This research found that young people responded to such experiences in a range of ways, including skipping school and responding with aggression, leading Penny Curtis (2008: 415) to observe that young people's 'emotional well-being was strongly influenced by their experiences of being obese'. She cautions that there is a need 'to be sensitive to the growing intolerance towards heavy people within schools and the judgements made by other young people on the bodies and social practices of those deemed to be overweight' (Curtis 2008: 417). Yet, the excessive bodies of overweight young people are not consistently spread across place or identities, and certain groups of young people are more or less likely to experience issues with their size according to the social groups they belong to.

To simply represent young people's bodies as marginalised, subcultural and excessive touches only on the margins of the experiences of young people's embodiment, identities and places. It might be useful to consider other approaches to young people's bodies and embodiment as issues about young people, place and identity are explored in the chapters that follow. Furthermore, the focus upon relatively negative aspects of young people's embodiment plays into stereotypes about young people as being deviant and in trouble, so I could be accused of further stigmatising young people given the content of this chapter. However, it also says something about the nature of academic research about young people, place and identity that the vast majority of work about young people's bodies focuses on those associated with the marginal, the spectacular and the carnivalesque. Future research could therefore usefully seek to explore the full complexity of young people's bodies and embodiment including reference to positive aspects and attributes associated with the places and identities of youthful bodies.

Key themes

- Young people's bodies are key sites for them to express their identities through dress, style or accessories.
- Young people's bodies are locations where members of society make judgements about who young people are, where they should be, and what they are doing.
- Research has tended to focus upon the marginal, subcultural and excessive aspects of young people's embodiment.

Project ideas

In what ways do young people have marginalised bodies, what are the processes of marginalisation in operation and how can these processes be challenged and overturned?

How useful is the concept of 'youth subcultures' for understanding young people's embodiment?

In what ways are young people's bodies regarded as excessive? What social factors contribute to these excesses and what are the advantages and disadvantages of such experiences?

Consider the experiences of young people who possess unwanted, persecuted or restricted bodies. In what ways do their bodies shape their experiences? How are their bodies shaped by others, and how might young people resist or contest such representations?

Suggested further reading

This is a brilliant collection of articles about the complexities of young people's bodies:

Colls, Rachel, and Hörschelmann, Kathrin (2010) *Contested bodies of childhood and youth*. Basingstoke: Palgrave.

This is a very useful introduction to the ways in which researchers have adopted a cultural focus to their work with young people:

Kehily, Mary Jane (2007) A cultural perspective. Kehily, Mary Jane (ed.) *Understanding youth: perspectives, identities and practices*. Sage: London. 11–44.

This collection offers a series of rich examples of the intersections between bodies, embodiment and space:

Kenworthy Teather, Elizabeth (1999) *Embodied geographies: space, bodies and rites of passage*. London: Routledge.

Excellent and accessible, this book explores understandings of the body and embodiment:

Longhurst, Robyn (2001) *Bodies: exploring fluid boundaries*. London: Routledge.

This is a fascinating exploration of young people's experiences of clubbing:

Malbon, Ben (1999) *Clubbing: dancing, ecstasy and vitality*. London: Routledge.

An excellent and accessible exploration of key themes in social geographies, focusing upon a range of scales:

Valentine, Gill (2001) *Social geographies*. Harlow: Pearson Education.

5 Home

Andrea: A place where I feel comfortable and that I belong to . . . (Holdsworth and Morgan 2005: 79)

Shannon: It's like a sanctuary . . . I think always it's a sanctuary, somewhere you feel at peace, somewhere you don't have to – you can just be yourself – where you feel welcome. (Henderson *et al.* 2007: 125)

Brad: . . . once I had my own place it was very much my place, it wasn't my parents, it wasn't me living with someone else. It was my house and to a certain extent . . . I felt 'this is my place and people live by my rules, I want to be at home in my own home'. And so it was during that first year [as a homeowner] that I came out. (Kenyon 2003: 110)

Youthful homespaces

For Andrea, Shannon and Brad, home is where they can be themselves and live according to their own desires. Within society, home is often represented as a positive place for people and a key site for how people construct their everyday lives (Blunt and Dowling 2006, Blunt and Varley 2004). 'Frequently, the home has been regarded as a place to withdraw to, a place of rest and a place where one has a large degree of control over what happens. It is seen as the place where individuals can assert their identities. Home then is understood as "something to be invested in, serving as it does as a locus for the identities of those who live there"' (Holloway and Hubbard 2001: 77). Also, unlike a number of other places, 'access to home is often restricted in different ways and so "home" is often understood as a place within which only certain people and things belong; it is a place to which a person or group of people can withdraw from the outside world' (Holloway and Hubbard 2001: 77). For young people then, home can be seen as an important site for the construction of their identities and a place where they can escape from the pressures or tensions sometimes associated with their lives outside home whether this is studying at school, college or university, volunteering or work, or even just hanging out with their peers.

Shelley Mallett quotes from the *Collins English Dictionary* definition of home (see Box 5.1), demonstrating the complexity, diversity and multiple meanings associated with home. Thinking specifically about the family home into which people are born, she observes that it:

> holds symbolic power as a formative dwelling place, a place of origin and return, a place from which to embark upon a journey. This house of dwelling accommodates home but home is not necessarily confined to this place. The boundaries of home seemingly extend beyond its walls to the neighbourhood, even the suburb, town or city.

Box 5.1

Home

1 The place or a place where one lives: *Have you no home to go to?*
2 a house or other dwelling
3 a family or other group living in a house or other place
4 a person's country, city, esp. viewed as a birthplace, a residence during one's early years, or a place dear to one
5 the environment or habitat of a person or animal
6 a place where something is invented, founded or developed: *the US is the home of baseball*
7 *a* a building or organization set up to care for orphans, the aged, etc. *b* an informal name for a mental home
12 *a home from home* a place other than one's own home where one can be at ease
14 *at home in, on, or with* familiar or conversant with
25 *bring home to. a* to make clear to. *b* to place the blame on

Collins English Dictionary (1979: 701), cited in Mallett (2004: 62–3)

> Home is a place, but it is also a space inhabited by family, people, things and belongings – a familiar, if not comfortable space where particular activities and relationships are lived.
>
> (Mallett 2004: 63)

Home therefore is not necessarily a single location or dwelling but can be associated with a number of different places and contexts in which young people live their lives. Moreover, Mallett comments that 'the home functions as a repository for complex, inter-related and at times contradictory socio-cultural ideas about people's relationship with one another, especially family, and with places, spaces and things' (2004: 84).

Somerville (1997) highlights how the meaning of home can include diverse representations such as physical security, legal ownership, investment, emotional security, self-expression and social status. Drawing upon different disciplinary perspectives, he identifies four main approaches to understandings of home (see Box 5.2).

Box 5.2

Theories and understandings of home

Territorial approaches – focus upon spatial boundaries, family territories and individual territories within the home. This emphasises issues of control, security and social relations at different scales (e.g. individual rooms, household, neighbourhood) often drawing attention to the physical, psychological and spatial realms.

Psychological and social-psychological approaches – focus upon the need for identity, control, privacy, security, intimacy and social status. This approach (which overlaps with territorial approaches) often draws attention to issues of attachment, connection and people's understandings and representations of home.

Phenomenological and developmental approaches – focus upon the continuity and stability of home across the lifecourse. Issues of value, ritual and individual and group interactions tend to be the focus of such approaches alongside themes of regulation and appropriation.

Sociological approaches – focus upon structure, agency and place. Issues of privacy, identity and familiarity draw attention to the interweaving of spatial, psychological and social relations within the home.

Adapted from Somerville (1997)

> All types of study have revealed the same recurrent meanings of home as the center of family life; a place of retreat, safety and relaxation, freedom and independence; self-expression and social status; a place of privacy; continuity and permanence; a financial asset; and a support for work and leisure activities.
>
> (Somerville 1997: 227–8)

Furthermore, attachment to home is likely to increase as length of residence increases, and previous research has demonstrated that men are more likely to see home in terms of achievement or status with women being more inclined to see it as a refuge or site of protection (Somerville 1997).

Although popular understandings of youth often focus on the public realm, those locations associated with privacy – such as the home – also hold a special place for young people and the construction and contestation of their social identities. I use the phrase 'special place' here, because, although home can be a loving, caring, open and transformative site for many young people (as suggested in the popular understanding outlined above), it can also be a site of resistance, aggression, anger and possibly even violence. Rachel Thomson (2007) outlines eight sets of critical moments in young people's biographies, three of which focus specifically on themes connecting with housing and home (see Box 5.3), thereby demonstrating the significance of home in young people's lives.

Furthermore, some young people experience life without home, or grow up in a setting where their connections to any sense of home – both in a symbolic and material manner – are very fragile, regularly destroyed and frequently absent. So, home can conjure up a complex mixture of emotions and memories for young people, depending on their particular experiences. As confirmed by Alison Blunt and Ann Varley (2004: 3), 'as a space of belonging and alienation, intimacy and violence, desire and fear, the home is invested with meanings, emotions, experiences and relationships that lie at the heart of human life.' It has been suggested that there has been 'an explosion of scholarship around the concept of home' (Jacobs and Smith 2008: 515) in recent years, but relatively little of this work focuses explicitly upon the experiences of young people. Yet, homes are both 'material and symbolic and are located on thresholds between memory and nostalgia for the past, everyday life in the present, and future dreams and fears' (Blunt and Varley 2004: 3). In an attempt to explore some of the diversity of

Box 5.3

Critical moments

Family

Being kicked out of home; parents splitting up; disclosing abuse; father remarrying; falling out with step-parents; parental unemployment; disowned by mother; one parent leaving the home; parental repartnering; reconciliation with estranged parent.

Death and illness

Death of a parent; aunt committing suicide; loss of a baby; diagnosis of dyslexia; diagnosis of chronic illness; death of grandparents; depression; falling ill; births; death of friend; death of mother, father, friends, grandparents, niece, aunt; grandmother moving away; rejection by mother; diagnosis of rheumatoid arthritis; near-death experience.

Moving

Moving town; moving house; moving country; siblings leaving home; becoming mobile – passing driving test; going to university.

Adapted from Thomson (2007: 101–2)

young people's experiences of home, in this chapter I focus upon four main themes: experiencing home, without home, leaving home, and making home.

Experiencing home

In terms of thinking about young people's experiences of home, David Sibley (1995: 129) notes:

> The home is one place where children are subject to controls by parents over the use of space and time and where the child attempts to carve out its own spaces and set its own times. The possibilities for conflict are considerable. Children may find the domestic regime oppressive because of rigid parental control of space, the availability of space in the home may limit opportunities for children to secure privacy, adults may feel that children get in the way and so on.

Although Sibley's comments are about childhood, they also say much about young people's experiences of home as they seek to manage the demands of parent(s) and create their own spaces and times. A comparison is also made between positional and personalising families, the former being about control and inflexibility and the latter about having a more equal balance of power between individuals in the home. Clearly, the nature of family relationships within the home often has powerful influences over young people's experiences. Furthermore, the nature of the space available to young people often changes during the course of the day as parents commonly determine what are adult spaces and adult times, creating a system that young people have to manage, negotiate and may often contest. An important aspect of young people's experiences of home is frequently about having their own space – or their own room – and as Sibley (1995: 113) observes this is often an issue which speaks of class differences and the presence of other siblings:

> Having one's own space is important in developing autonomy and this distinguishes the middle-class child who is part of a small family from one with many siblings or living in poverty. Particularly when a child has been given its own bedroom, then the space may be appropriated, transformed and the boundaries secured by marking that space as its own.

So, young people's experiences of home, and the way in which they articulate and develop their identities at home, often depend upon the intersection of factors such as the existence of a positional or personalising family situation, the rigidity of the use of space set down by adults, the extent to which the young person has their own room, as well as the presence of other siblings or family members in determining the use, allocation and management of space.

Space within the home is important, as is the nature of relationships with other people living there. As Martin Robb (2007: 314) observes:

> [Y]oung people's experience of family life is extremely diverse. While a majority may live with two biological parents, others live with a lone parent, or have a close relationship with a non-resident parent or step-parent. For some young people, such as those in care or custody or lacking support at home, relationships with key workers, teachers or youth workers may be more significant.

So, the other adults and the nature of young people's relationships with them often have a profound influence over what happens at the intersection of age, identity and place. Furthermore, young people are increasingly choosing to live in the parental home for longer periods, highlighting the importance of experiences of home as a crucial topic (Coles *et al.* 1999). Furthermore, the spaces within the home and how these are used, controlled and shaped by young people have important influences over how young people construct and contest their identities. Sian Lincoln conducted research with teenagers about bedroom culture and music. She found that music was a key mechanism by which young people gave meaning to their bedroom spaces and had a sense of control and autonomy in such locations. Lincoln (2005) also found that young people were increasingly using the stereo system in collaboration with the internet to enhance their musical experiences, thereby highlighting how the private location of the bedroom is connected to different public domains (see also Hodkinson and Lincoln 2008). In addition to the use of the internet, young people may also engage with other forms of computer technology and practices (Donovan and Katz 2009). Ian Shaw and Barney Warf (2009: 1332) discuss the virtual worlds of the video games and how they 'invite players to explore, consume and experience virtual universes.' In particular, they discuss the ways in which video games reinforce assumptions about gender, race and Orientalist others through the depiction of characters and images present within video games.

Bob Coles, Julie Rugg and Jenny Seavers (1999) identify six groups of young people who live at home. First, there are the 'pre-decision stayers' who are those young people who have not thought about their housing situation and are living in the parental home. The second group – 'willing stayers' – represents those young people who have consciously decided to stay at home for a variety of reasons including cost, family relations and a general preference for staying at home. 'Reluctant stayers' form the third group identified and this includes young people who want to move out but are unable to do so financially. Fourth are the 'reluctant returners' who have returned home typically due to their inability to sustain independent tenancy. The 'willing returners' represent the fifth group and 'students studying away from home' the sixth. The set of six 'groups' demonstrates the diversity within young people's experiences of living at home, and indeed, even within each category, there are a number of reasons and explanations for young people's motivations to stay at home or not.

Young people who identify as lesbian, gay, bisexual, transgender or queer often experience the difficult decision about whether or not to come out at home and, if so, when they do this. Here,

> the overwhelming and taken for granted heterosexuality of the family home can be experienced as oppressive and alienating . . . young people can feel guilt and discomfort at concealing their sexuality from family members. This can serve to drive a wedge between young people and their parents.
>
> (McNamee *et al.* 2003: 125)

So, although there is the possibility that coming out may lead to the creation of a supportive home situation, it could also risk being an alienating and destructive process. This experience is also bound up in expectations around social identities – such as gender, class and family expectations. Furthermore, although young people are often represented as less competent than adults, for some non-heterosexual young people, their performing of a heterosexual identity at home is often about protecting the emotional stability of their vulnerable parents:

> Young people are often presented as emotionally immature, less competent than their parents and in need of protection from the turbulence of adult relationships. Yet, many of the young people who participated in this research represented themselves as more worldly than their parents, arguing that they passed as heterosexual in the family home in order to protect their emotionally vulnerable parents from the hurt and disappointment of discovering their sexuality, and from having to make similar choices about whether to disclose this information to other relatives and their friends.
>
> (McNamee *et al.* 2003: 127).

Furthermore, these experiences are often associated with appropriate gendered behaviour and so negotiating sexuality in the family home is often different for young men and young women as their sexual and gendered identities are constructed in relation to – and often against – those of their parents. Box 5.4 discusses the process of young gay men coming out to their heterosexual fathers, highlighting the ways in which ideas about sexuality and masculinity are bound up in such practices.

Without home

Young people's negative experiences of home such as conflict within the family home – or other place where young people might be living – can lead to a situation where they become homeless, and so are

Box 5.4

Young gay men coming out to their heterosexual fathers

Research has explored the experiences of young men who 'come out' to their heterosexual fathers, the intersections of sexuality and masculinity in this process, and how this relates to constructions of home (Skelton and Valentine 2005). This work discusses three cases of young men 'coming out' to their fathers and the various responses of rejection, condemnation, denial and acceptance exercised by the fathers, as well as the ways in which new relationships were established (or not). This work also touches on the context of 'home' and how this changes through these processes and interactions.

The first example is Robbie, a 22-year-old graduate student who described his parents as 'conservative, old in their ways and very strict' (p. 211). He came out to his brother who was very supportive and then decided to tell his parents. His mother was insulting, upset and hysterical, whilst his father condemned him as being a 'paedophile' and an 'abomination'. As Tracey Skelton and Gill Valentine (p. 212) note 'Robbie is the very embodiment of the "other" which the father has clearly rejected in order to construct his own masculine identity.' However, further issues arose when Robbie and his boyfriend of two years were invited to Robbie's brother's wedding. Robbie's parents refused to turn up if Robbie insisted on bringing Philip to the wedding. In the end, Robbie did not take Philip to the wedding and so his father had succeeded in ensuring that his homophobic hegemonic masculinity was not threatened by the homosexuality of his son.

A rather different situation was experienced by Mark, a 19-year-old young man employed in a security firm. He was brought up in a supportive household and felt uneasy about not revealing his sexual identity to his parents. He was now sexually active but was worried that he had contracted HIV so decided to go for a test and took some time off work. His father questioned him about not being at work, and Mark's evasive response made his father suspicious so he kept questioning him and Mark explained about the test. His father was very angry and called a family meeting – the first in the family's history – in order to set down ground rules for how Mark should behave at home (p. 215):

> . . . they just told me how it was gonna be . . . they laid down these five golden rules . . . I wasn't to flaunt my homosexuality in the house . . . I wasn't to talk about that side of my life . . . I wasn't to bring any boyfriends home . . . I wasn't to bring any of my friends who were 'that way' home . . . they decided I couldn't tell anybody else.

Although the situation was very strained for some time, Mark's dad had met his boyfriend a couple of times, invited him to the New Year's Eve party and was helping them to move into their own home together, which Mark and his boyfriend were in the process of setting up. His father's heterosexual masculinity was very gradually becoming slightly more accepting of Mark's sexual identity.

A third and final example is the case of 21-year-old Noel who lived with his mother and stepfather. Although he had come out to friends at college, it was only during an argument at home that he disclosed his sexuality. Although his parents initially rejected this as a phase, they were generally rather bemused about the whole situation and his 'real dad' was accepting of this when he found out. It was, instead, Noel's sister who was most hostile about his sexual identity. In a rather different scenario from that experienced by Robbie and Mark, Noel found that his step-father ended up displaying a tolerance and 'genuine interest' (p. 218) about Noel's sexuality, and the acceptance he received from both fathers was extremely important to him.

Adapted from Skelton and Valentine (2005)

effectively without a home. There are over 100 million homeless young people in the world (Rosenthal and Rotheram-Borus 2005) which raises important questions about young people's health and well-being and experiences of marginalisation alongside practical questions about service provision for homeless youth and suitable accommodation to meet the needs of young people. Being labelled 'homeless' stigmatises young people, constructing them as delinquents, associating them with suspicion and correlating them with negative behaviours and places (Pain and Francis 2004, Ruddick 1996). Abby Peterson's (2000) work in Sweden with young women who had experienced homelessness discusses the ways in which these young people were made to feel abnormal and constructed as outsiders due to the existence of a belief that there were no homeless people in Sweden. Furthermore, the label 'homeless' applies only to young people's housing circumstances so only speaks to one aspect of their lives and, as such, they are likely to be a very diverse group yet are often not labelled as such (Pain and Francis 2004).

Joe Oldman confirms that 'many young homeless people are forced to leave the parental home by a variety of "push factors". These may include violence, physical and sexual abuse, eviction, family breakdown, and arguments with step parents.' Furthermore, these 'push factors' are also often combined with a lack of affordable housing, rising unemployment and low pay heightened by the way in which 'homelessness creates a vicious circle of no home no job, no job no home' (Oldman 2004: 114–15). This means that homeless young people are one of the most vulnerable groups in society, and are specifically at greater risk than young people living at home – 'greater risk of psychiatric distress, unemployment, suicidal acts, pregnancy, survival sex, dropping out of school, involved with the criminal justice system or substance abuse' (Mallett *et al.* 2004: 338). Furthermore, they often occupy hidden, marginalised and invisible locations and so their experiences of exclusion are heightened further (Pain and Francis 2004).

The experiences of homeless young people are diverse and vary widely depending on their experiences of entering homelessness, their strategies for managing their circumstances, the standard of any assistance they might receive and the extent to which they use this assistance. Drawing upon quantitative research conducted about homeless young people in Australia and the USA, Shelley Mallet *et al.* (2004) found that there were four main groups of homeless young people: partnered homeless; socially engaged homeless, service connected-harm avoidant homeless; and transgressive homeless. The first group represented around 15 per cent of the sample and tended to spend most of the time with their partner, were more likely to be female and were more likely to be part of the Australian rather than American sample. The second group – the socially engaged homeless – tended to be older, more likely to be male and Australian. They were often involved in a range of positive and negative activities in using up their time and were the most engaged of the four groups identified. The service connected-harm avoidant group were the largest in the sample (44.6 per cent) and were also the youngest of the four groups and tended to be female. As suggested by their name, this group tended to spend most of their time being connected in a variety of levels to available services. The final group – transgressive homeless – made up nearly 19 per cent of the sample and included older and male young people, as well as those who had been away from home the longest. They were also more likely to be American then Australian. This study demonstrated connections and differences between experiences of homeless young people in two different countries, as well as the ways in which the young people's experiences of homelessness are often relational to other factors such as whether or not they have a partner and the availability of appropriate services.

The way in which homeless young people's experiences are linked into broader global processes has been highlighted by Ruddick's (1996) work with homeless young people in Hollywood. She relates the experiences of homeless young people in terms of how constructions of identities as young people and as homeless interconnect with different places and constructions of space. Her work explores the development of a punk street culture from the 1970s through to the 1990s; in particular she demonstrates how the use and appropriation of different places results in punks having the upper hand, thereby recreating the meanings and associations of different city spaces. They squatted in derelict mansions and apartments in Los Angeles and Ruddick (1996) argues that they displayed sophisticated insights into the workings of capitalism and the outcomes of globalisation as well as showing the tenacity to resist particular institutional mechanisms through confronting and contextualising their marginality through occupying public space.

The situation in Britain is equally challenging for homeless young people, where it is estimated that, in 2002, 250,000 young people aged 17–25 were living in temporary accommodation (Pain and Francis 2004). In this context, 'the main factor leading to youth homelessness is family conflict, which may result in young people choosing or being forced to leave the parental home' (Pain and Francis 2004: 97). However, as suggested above, this situation is further complicated by social, political and economic factors such as the high level of youth homelessness

> has been exacerbated since the 1980s by a number of structural, economic and policy factors, including high youth unemployment, a reduction in state benefits (especially for 16–18 year olds, but also for 18–25 year olds), and a lack of easily accessible cheap accommodation owing to the lack of new build, the sale of council stock, and low levels of cheap privately rented or Housing Association accommodation.
>
> (Pain and Francis 2004: 97)

Homeless young people therefore occupy an ambiguous and contested place on the boundaries between public and private as they lack the sense of privacy that is available to many other young people, spend the vast majority of their time in public places, yet simultaneously many such locations are constructed in particular ways, as noted above, that work to marginalise and stigmatise the place and those people inhabiting it. Furthermore, as Rachel Pain and Peter Francis (2004: 102) observe:

> Homelessness is a dimension of risk additional to those of gender, race, sexual orientation, age and able-bodiedness; one which has a

> compounding effect on the other inequalities which excluded groups experience. It emerged that the greatest risk for homeless young people is not the street, but the homes they have left behind, the relationships young women have with men, and young people's contact with the police.

So, being labelled with the identity of homeless works to open young people up to an additional set of risks alongside those already associated with their other social identities. Furthermore, witnessing a reversal of popular understandings of home, many homeless young people find that one of the riskiest places is the home they have left behind.

A common stereotype of homeless young people is that they are involved in various forms of criminal activity. From their research in England, Rachel Pain and Peter Francis (2004: 107) found that most of the criminal activity that homeless young people were involved with was 'petty crime or "lifestyle" crime such as begging or drug use', and that various forms of offending became normalised forms of behaviour in order to get by. Yet, challenging commonsense assumptions about this social group, they also found that 'homeless young people are more at risk from crime victimisation, before and while they are homeless, and also more likely to suffer multiple and repeat victimisation' (Pain and Francis 2004: 102), especially harassment and violence from the police. Many young men reported being searched by the police regularly, and young women reported violence from partners when homelessness was a regular experience.

Leaving home

Leaving home is often regarded as a key phase – or a critical moment (see Box 5.3) – in young people's transitions towards independence and adulthood and is often a complex process rather than a one-off event (Image 5.1). Furthermore, the point at which young people leave home varies considerably across different Western nations and, on average, women tend to leave earlier than men (Holdsworth and Morgan 2005). As Gill Jones (1995: 62) confirms, leaving home

> is often only one stage in a process which eventually leads to household formation and independent housing. This process can take several years, involving two or more leaving home events. Many young people move into transitional intermediate households, from where they may either move on into independent households of their

Image 5.1 Leaving home.

> own, or return to the family home again. The transition out of the family home is thus a complex process which can be reversible, involving returns as well as departures.

Furthermore, this process is 'overlaid with emotional and moral overtones' (Jones 1995: 143) which tend to result in it being classified as either normal or strange depending on factors such as family expectations, the age of the young person and the nature of the local housing market.

The complex process of leaving home often involves returning home for a number of reasons:

> Many young people do return home: this may be because it is convenient for them to do so, because they have missed their families, because the reason they left has passed, or because they decide the time is not yet right for them to make the final break from the parental home.
>
> (Jones 1995: 65)

In her influential research about leaving home, Gill Jones (1995: 91) identifies four different groups of young people: supported returners, unsupported returners, supported independence and unsupported independence. Although both the first and third groups received help from parents in setting up home, for a variety of reasons, those in the first group returned home and those in the third didn't. With the second and fourth group, parental support was absent, yet the second group returned home and the fourth group did not. However, although there was much diversity in terms of whether or not young people returned home after leaving and the extent to which they received any help from their parents in leaving home, so too the process of leaving was often complex and not simply a singular transition as young people often moved to an intermediate home before becoming fully independent:

> People tend to move from the *parental home* into independent households, often via an *intermediate household* stage. Intermediate households are ones in which the young person may be living in what could be a considered a 'surrogate parental home', for example with relatives, or in group housing such as a barracks . . . in which neither they nor their peers are the heads of households. The independent households may be partnership homes, but it appears that young people increasingly set up home alone or with their peers, sometimes as a precursor to partnership living (and thus part of the extended current process of transition to adulthood), and these young people appear to be far more emancipated from parental authority than those in intermediate households.
>
> (Jones 1995: 109)

Alongside acknowledging the diversity of destinations and experiences of young people leaving home, it is also important to recognise differences across countries of the mean age of home leavers (see Table 5.1). As Clare Holdsworth (2000: 202) notes:

> [I]n Northern Europe and North America, young men and women from higher socio-economic backgrounds are more likely to leave home earlier for independent living, while leaving home at older ages to establish a family household is more common among working-class young people, especially women.

An example of some of these differences is highlighted by research which uses data from the National Child Development Study in the UK and the Spanish Sociodemographic Survey. Here, Clare Holdsworth (2000) explores patterns of leaving home for young people in Britain and Spain. For respondents who left home before 1981, the mean age of leaving home was 22 and 21 for men and women respectively

Table 5.1 ***Comparison of median age for leaving home***

Country	Median age (men)	Median age (women)
Europe		
Austria	21.8	19.9
Belgium	23.3	21.5
Czech Republic	23.8	21.2
Germany (east)	22.4	20.6
Germany (west)	22.4	20.8
Finland	21.7	19.8
France	21.5	19.8
Hungary	24.8	21.3
Italy	26.7	23.6
Latvia	24.1	21.3
Lithuania	20.3	19.8
Netherlands	22.5	20.5
Norway	21.4	19.8
Poland	25.8	22.5
Portugal	24.3	21.8
Slovenia	20.9	20.5
Spain	25.7	22.9
Sweden	20.2	18.6
UK	22.4	20.3
North America/Australia		
Australia	18.6	18.1
Canada	22.7	20.9
USA	21.5	20.8
Asia		
China	25.0	23.9
Japan	22.4	23.8
Malaysia	24	22
South Korea	26.6	23.5

Adapted from Holdsworth and Morgan (2005: 8–9)

in Britain and 24 and 23 respectively in Spain. Furthermore, in 1991, 17 per cent of men and 13 per cent of women in Spain were living in the parental home with the British figures showing that 9 per cent and 5 per cent respectively of men and women were living in the parental home. Holdsworth found that over 70 per cent of young people in Spain left home for partnership, whilst in Britain this was less that 60 per cent for women and only 49 per cent for men. This work has also shown that parents' education influences the age of leaving home more than the occupation of the father does.

The act of leaving home not only influences the personal, social and economic situations of young people but also has a direct influence on familial relations more broadly (Holdsworth 2000).

Drawing upon her research about young people leaving home in Britain, Spain and Norway, Holdsworth (2007) has discussed the significance of intergenerational relations in such contexts. In particular, she explores the gendering of intergenerational relations focusing upon the experiences of three young women and their mothers. She found that there were strong senses of dependency between young women and their mothers and that this was often a two-way process and something that both mothers and daughters identified with in a positive way. For the young women, it was through their responsibilities at home that they were able to develop independence and move towards adulthood. Their connections with their mothers were also reinforced through helping each other in times of need, caring for each other and providing general support and companionship (Holdsworth 2007).

Making home

Generally, research about young people, housing and home tends to focus on key transitions, such as the process of leaving home for university or to set up a family. A very important aspect of young people's lives is often the experience of 'making home' and the practices and processes associated with this. Whether this is about getting a mortgage and becoming an owner-occupier, buying a house with a friend, renting from the council or the negotiation of other housing options, it can be a life-changing and exciting time for young people, as well as an experience that can be associated with stress, anxiety and insecurity. The processes and methods whereby young people set up home have become increasingly diverse as a result of changes in housing markets, different trends within partnership formation, changing youth employment opportunities and the increase in further and higher education provision. There is not much research on this topic as a result of the ways in which young people and housing tend to be stereotyped:

> [F]ew people would deny that housing is centrally important in the lives of young people, yet young people, particularly those under 18 years, remain either a muted group or are regarded in simplistic terms as 'a housing and homelessness problem' in housing policy research.
>
> (Rowlands and Gurney 2000: 121)

An important issue about 'making home' is about the ways in which ideas about housing situations are bound up with social norms and family expectations. For example, an important question is when is the right time for a young person to leave home and set up on their own (Holdsworth and Morgan 2005), and often preferences about tenure are 'culturally mediated and socially constructed' (Rowlands and Gurney 2000: 122), relying on intergenerational trends in knowledge and understanding about housing.

Important research conducted by Rob Rowlands and Craig Gurney (2000: 126) found that 16- and 17-year-olds see housing as part of a 'package' of goods that they aim to collect in their desire for social mobility, as there were issues of 'image and prestige' associated with housing. Owner occupation is strongly associated with success whilst council housing is connected with failure. Challenging popular notions of young people's understandings of housing, they found:

> The young people who contributed to this research displayed an informed understanding of the housing system and strong opinions about the relative merits of renting or owning in contemporary UK society. This understanding appeared to be built on information from parents, friends and the media.
>
> (Rowlands and Gurney 2000: 127)

Although young people may see owner occupation as being associated with success, for many this is becoming an increasingly challenging status to achieve as a result of the social and economic changes mentioned above. Janet Ford (1999) has suggested that one reason for the declining number of young people seeking owner occupation is due to changing attitudes about housing as a result of a number of factors such as their employment situation, the cost of property, the need for a deposit, the commitment to a mortgage, concerns about coping with mortgage payments alongside the flexibility in renting.

Elizabeth Kenyon notes 'increasing numbers of 20- and 30-something young adults in the UK are rejecting early partnership formation, and are instead living alone or in multi-adult shared households unconnected by marriage, co-habitation or family ties' (Kenyon 2003: 103, see also Heath and Kenyon 2001, Kenyon and Heath 2001). Similar trends are also evident in the USA and northern Europe (see Box 5.5). Young people living in shared housing are often regarded 'as experiencing an extended transitional period of "making do", where

Box 5.5

Home alone

Thinking about being 'home alone' may provoke anxiety when thinking about younger children, yet for young people, solo living is becoming increasingly popular. 'Solo living is an emergent feature of young adult lives, with 11 per cent of men and 6.5 per cent of women aged 25–29 living alone at the time of the 1991 Census in the UK (Chandler *et al.* 2004). Wasoff *et al.* (2005) note that 5 per cent of men and women aged 16–24 were living alone in 2002, with 16 and 8 per cent of men and women respectively aged 25–44 also living alone. Most young people who live alone have never married with a small proportion having separated from a partner. Solo living for men is common at either end of the social class scale whereas for women it is closely associated with professional status (Chandler *et al.* 2004).

Adapted from Chandler *et al.* (2004) and Wasoff *et al.* (2005)

adulthood is placed "on hold" until a truly independent adult home can be established' (Kenyon 2003: 103), and so there is a strong association between adulthood and independent housing status, with any other housing situation being seen as being representative of a young person in a transitional phase.

Drawing upon research in the USA, William Clark and Clara Mulder (2000) have explored young people experiencing their first move after having completed education. They found that the vast majority rent initially with only around 10–15 per cent entering home ownership immediately. Although renting is still the most popular for both young couples and single people, home ownership is more likely for young couples than singles. Furthermore, young people whose parents own their own home are likely to become home owners too. Older young people are less likely to share than younger people. Furthermore, 'the first destination on the housing market after nest leaving is to a certain extent predictive for the further course of the housing career' (Clark and Mulder 2000: 1657).

Key themes

- Home has multiple meanings for young people and the construction and contestation of their identities, as it can be associated with happiness and contentment as well as violence and abuse.
- Experiences of home and other events that influence how home is experienced (such as a parent or sibling leaving) are often seen as 'critical moments' in young people's home lives and in their transitions to adulthood.
- The form of adult control at home, the presence of other siblings and whether or not a young person has their own room are important factors in a young person's experiences of home.
- A number of young people are without a home and are increasingly stigmatised as a result of this. They often occupy invisible places on the margins of society and regularly experience harassment and victimisation.
- Leaving home is a complex process for young people and often involves return moves and moves into intermediate households before a fully independent housing situation is reached.
- Young people's experiences of 'making home' are influenced by family expectations, intergenerational knowledge about housing and the social aspirations of young people themselves, many of whom are knowledgeable about their housing situations and opportunities. Increasing numbers of young people are living alone or in shared housing situations.

Project ideas

Discuss the ways in which the multiple identities of young people shape and are shaped by their experiences of home as simultaneously empowering and oppressive.

What type of household formation are you currently living in, how does this make you feel about and understand home and how have your experiences changed through time?

What are some of the consequences of the increasingly diverse nature of household living arrangements for society, the economy and politics?

In what ways do the experiences of young people without a home work to reshape what is understood by the concept of home?

Suggested further reading

This books offers a fascinating exploration of the multiple meanings and understandings of home:

Blunt, Alison and Dowling, Robyn (2006) *Home*. London: Routledge.

This is an excellent account of young people's experiences of leaving home in the context of the transition to adulthood:

Holdsworth, Clare and Morgan, David (2005) *Transitions in context: leaving home, independence and adulthood*. Berkshire: Open University Press:

This excellent article details the ways in which young people use music in their bedrooms to express their identities:

Lincoln, Sian (2005) Feeling the noise: teenagers, bedrooms and music. *Leisure Studies* 24(4) 399–414.

This article offers an interesting account of young people's views about housing and home ownership:

Rowlands, Rob and Gurney, Craig M. (2000) Young people's perceptions of housing tenure: a case study in the socialization of tenure prejudice. *Housing, Theory and Society* 17(3) 121–30.

This is a rich social and political account of young people's experiences of homelessness:

Ruddick, Sue (1996) *Young and homeless in Hollywood: mapping social identities*. London: Routledge.

This is a very useful exploration of young people's experiences of housing and home in the context of social policy:

Rugg, Julie (1999) *Young people, housing and social policy*. London: Routledge.

6 Neighbourhood and community

Jasmine: 'What I think of when thinking of my neighbourhood is Latin music and hip hop playing in the street. In the summer people hang out on the corner and in the park and what seems to be thousands of children running around playing on the street. Sprinklers on, with the smell of BBQ in the distance. The icy man yelling icy's coco and cherry, the ice cream truck playing in the distance. (Cahill 2007a: 202–3)

Paul: Well, when you get into trouble it has a lot to do with your background and where you live. If you live in a dodgy area you are bound to get into trouble. You meet with the wrong people. If you knock about with the wrong people you have had it. It's the areas as well . . . there's nothing to do here. You start pinching and you get caught. (Shildrick 2006: 68)

Anna: I would pray but I don't say I'm religious, I wouldn't read the bible, I couldn't tell you the ten commandments but I do believe in God, I just think it's the way people my generation is, it's like my granny said to me she doesn't know what's going to happen when her generation goes because there's nobody goes to church and she's true but the only time our church is ever filled is whenever is Christmas Eve or when there's a christening or something. (Henderson *et al.* 2007: 116)

Neighbourhoods and communities of youth

Jasmine, Paul and Anna are all talking in different ways about how neighbourhood and community are important parts of their lives as young people. Growing up in New York City, Jasmine describes the music, culture and activities of her Lower East Side neighbourhood. Paul reflects upon how his background and neighbourhood interrelate with who he hangs around with and the activities he engages with, and Anna talks about her sense of being a part of a religious community and how this varies compared with other generations. Encapsulating the street and public space, neighbourhood and community are often very important to young people, especially those who have restricted mobility due to being too young to drive or who are economically marginalised and therefore unable to afford a car and restricted in their ability to pay to travel on public transport. The spaces of the neighbourhood and the community therefore become the primary domain of some young people's lives given the extensive amount of time they spend in such locales. This is the case for many people, as Kintrea *et al.* (2008: 12) observe: 'even though many people are increasingly mobile across urban space, immediate neighbourhoods remain a significant factor in people's lives.'

The concept of community is one of the most contested in the social sciences. Often used interchangeably with 'neighbourhood', 'community' can be characterised in three ways: '(1) common interests between people, (2) a common environment and locality, (3) a common social system or structure' (Henderson 2007: 127). In seeking to define the neighbourhood, Henderson (2007: 127) writes:

> At an elementary level, neighbourhood can be described as the sphere of the local, the place where you grew up or currently live. It can refer to the few streets that border your house or it may signal a more broadly understood meaning, conjuring up an image of a particular locality such as that estate, that side of town, or that part of the city.

Kearns and Parkinson (2001) complicate the definition of neighbourhood further by demonstrating the multiple scales of neighbourhood (Table 6.1). For them, the home area is about five or ten minutes' walk from home and is a crucial aspect of everyday life in terms of psycho-social well-being. It is important in terms of how relaxed people feel, how they connect with others and is crucial in fostering a sense of belonging. Overall the neighbourhood and

Table 6.1 ***Scales of neighbourhood***

Scale	*Predominant function*	*Mechanism(s)*
Home area	Psycho-social benefits (for example, identity, belonging)	Familiarity Community
Locality	Residential activities Social status and position	Planning Service provision (e.g. youth clubs) Housing market Employment
Urban district or region	Landscape of social and economic connections and opportunities	Leisure interests Social networks

Adapted from Kearns and Parkinson (2001: 2104)

community can be defined in both symbolic and physical ways but the focus tends to be on the connections, similarities and relationalities that create a common sense of neighbourhood and community, as well as differences and disjunctures between different groups of people that operate to highlight differences in neighbourhood and community affiliations (Tables 6.1 and 6.2).

In terms of neighbourhood and community, the dominant representation of young people is of figures that are disruptive and almost always viewed in a negative light. Yet, 'young people are often highly aware of negative adult attitudes towards them and this contributes to a sharply different experience of and purchase on their communities from that of an adult population' (Henderson 2007: 124). Furthermore, some neighbourhoods have specific plans or policies focused upon their regeneration or use, dictating the extent to which young people may benefit or not from the allocation of resources or the use of various neighbourhood places. In such circumstances, the ways in which young people construct and contest their identities can be shaped by local neighbourhood, community and urban policies and practices.

Neighbourhoods and territorialism

For many young people, their sense of identity and community is shaped by where they live, the territorial affiliation they hold and the tensions that exist between them and groups of young people from neighbouring communities (Image 6.1). 'While these territorial divisions do not appear on any town map, they are drawn in mental maps and are often a visible part of the streetscape' (Henderson *et al.* 2007: 63). Kintrea *et al.* (2008: 4) argue that territoriality is defined as 'a social system through which control is claimed by one group over a

Table 6.2 ***Indicators of community quality from the perspectives of young people***

Positive indicators	*Negative indicators*
Social integration: young people feel welcome and valued in their community	Social exclusion: young people feel unwelcome and harassed in their community
Cohesive community identity: the community has clear geographic boundaries and a positive identity that is expressed through activities like art and festivals	Stigma: residents feel stigmatised for living in a place associated with poverty and discrimination
Tradition of self-help: residents are building their community through mutual aid organisations and progressive local improvements	Violence and crime: due to community violence and crime, young people are afraid to move about outdoors or move outside their immediate local area
Safety and free movement: young people feel that they can count on adult protection and range safely within their local area	Heavy traffic: the streets are taken over by dangerous traffic with non-road users excluded
Peer gathering places: there are safe and accessible places where friends can meet	Lack of gathering places: young people lack places where they can meet safely and play with friends
Varied activity settings: young people can shop, explore, play sports and follow up on other personal interests in the environment	Lack of varied activity settings: the environment is barren and isolating, with a lack of interesting places to visit and things to do
Safe green spaces: safe, green spaces with trees, whether formal or wild, extensive or small, are highly valued when available	Boredom: young people express high levels of boredom and alienation
Provision for basic needs: basic services are provided such as food, water, electricity, medical care and sanitation	Rubbish and litter: young people read rubbish and litter in their environment as signs of adult neglect for where they live
Security of tenure: family members have legal rights over the properties they inhabit either through ownership or secure rental agreements	Insecure tenure: young people, like their parents, suffer anxiety generated by fear of eviction, which discourages investment in better living conditions
	Political powerlessness: young people and their families feel powerless to improve conditions or influence political processes or policies

Adapted from Chawla (2002: 229)

defined geographical area and defended against others.' Drawing upon research about 'anti-territorial' projects in Bradford, Bristol, Glasgow, Peterborough, Sunderland and Tower Hamlets, their research found that territoriality was a key aspect of everyday life for young people living in each city, and was often passed down the generations, emphasising its historical nature:

> Territoriality was often connected to very rigid geographical borders between neighbourhoods and communities and was often connected back through the generations. Territoriality and territorial behaviour is regarded as being transmitted through generational interactions as stories are passed down through families. The limited opportunities of some young people also work to intensify such accounts.
>
> (Kintrea *et al.* 2008: 5)

Image 6.1 Territorialism.

Young men aged 13–17 were most likely to be involved in such behaviour, although so too were men in their twenties. 'Young people often had positive motivations, such as developing their identity and friendships, for becoming involved in territorial behaviour, but territorial identities were frequently expressed in violent conflict with territorial groups from other areas' (Kintrea 2008: 4).

Furthermore, this research found that territoriality tended to be stronger in disadvantaged neighbourhoods and communities, where limited opportunities, restricted horizons and possibly even difficult family circumstances shaped young people's engagement with their local area. Ethnic divisions within cities often heightened such territorial behaviour, resulting in constrained mobility, limited access to services and increased risk of violent assault. The strength of such boundaries and restrictions left many young people feeling unable to cross into particular neighbourhoods or communities, particularly in the evenings and at weekends. Fear of violence for some young people also made such negotiations particularly tense. These divisions often existed for young people involved in group territorial behaviour as well as other young people who happen to live in a particular community but end up being in the wrong place at the wrong time. As such, the territorial behaviour of a few often resulted in the majority of young people being stigmatised and associated with violent and aggressive behaviour. Kintrea *et al.* (2008) also found that the territoriality was long-standing and generational, as stories, accounts, behaviours and attitudes were passed down through the generations. Territorial behaviour was not simply confined to teenagers, although Kintrea *et al.* (2008: 27) note that it was striking that there was a 'very strong level of identity with neighbourhood held by young people as they entered their early teens.' Often, conflict was connected with ethnic group membership, although there were also accounts of conflict between and within different ethnic groups. From their research, they developed a list of factors that motivate young people to be involved in territorial behaviour (Box 6.1).

Often territorial behaviours and attitudes are intersected by other forms of affiliation and identification, such as those associated with politics, religion, ethnicity or the state. Through work in Northern Ireland, Madeleine Leonard (2006) has demonstrated the ways in which young

Box 6.1

The motivations to be involved in territorial behaviour

Territorial friendship networks and groups provide an alternative to connections with family and home.
Respect and recognition gained within the peer group and local community.
A sense of ownership over place and a desire to protect the local area.
Crossing boundaries could be seen as insulting and lead to conflict.
The protection and ownership of young women by young men.
A form of leisure activity or 'recreational violence'.
Occasionally connected with material crime or financial gain.

Adapted from Kintrea *et al.* (2008: 6)

people's territorial behaviour is motivated by political and religious territories in the context of sectarianism. 'On each side of the peace walls, symbols of sectarian identity are at their height being manifested in political murals and houses and streets decorated with Irish or British flags' (Leonard 2006: 229). Children and young people engaged with these interface areas and sites through making their own marks on walls, through their restricted use of different neighbourhood spaces and through claiming strips of land situated in parks that overlapped the divisions. Some children tried to conceal their religious identities when visiting leisure centres and downplayed their support for football teams that might work to associate them with particular religious persuasions or spaces. In terms of their use of a cinema and shopping complex near the city centre, Leonard found that Catholic and Protestant teenagers used different entrances to the complex knowing that the entrance they used resulted in assumptions being made about their religious identity. Furthermore, groups of young people often spent time hanging around each entrance in order to claim their ownership of this territory. As such, 'leisure spaces were ways of asserting group allegiance' and the 'controlling of territory was part of this process' (Leonard 2006: 231). Plus, despite the harshness of the strict territorial divisions within their everyday environments, Leonard found that the 14–15-year-olds she worked with were able to transcend these issues by focusing on 'wider geographical features of nature such as the rivers and mountains that surround the wider environment.' This demonstrates young people's agency and ability to see beyond the harsh divisions and territories present in their everyday lives.

In his work with young people from a range of ethnic backgrounds living in two neighbourhoods in London, Les Back (1996) draws attention to the multiple ways in which young people jostle over the use of neighbourhood space. He found that young people talk about issues of race and racism through the language of neighbourhood, community and territory. Within the multi-ethnic neighbourhood of Southgate, Back (1996) found two contrasting ways in which young people constructed their sense of community. First, some young people adopted the 'harmony discourse' by arguing that their local area is free of racism and that all people are living in harmony. Both black and white young people adopted such a strategy. The second approach, which Back (1996) refers to as the 'black community discourse', is one which characterises the neighbourhood as being a site of political action for black people in their struggle for recognition within different institutions and within society in general. Both approaches were often adopted alongside each other or used interchangeably by young people.

Back (2002) has also argued that the territorial practices and behaviours of the young people involved in his research resulted in the reinforcements of particular stereotypes associated with black masculinity. Young white men and women often regarded young black men as 'prestigious figures' (Back 2002: 39) within the local community, drawing upon assumptions about the forms of culture and style adopted by black youth. 'Yet at the same time black young men could be characterised by whites as undesirable, violent, dangerous and aggressive' (Back 2000: 39), thereby highlighting the often contradictory, racialised and gendered constructions associated with young black men. The young black men in the research argued that the white areas were dangerous places that were best avoided (Back 2000). Alongside this, Vietnamese young people tended to be viewed as vulnerable, soft and effeminate. Back (2005: 19) therefore argues that 'racism is by nature a spatial and territorial form of power.'

Neighbourhoods and poverty

As well as being divided along political and religious lines, neighbourhoods and communities are often separated along the lines of wealth and social class position. This is often heightened by the tendency of middle- and upper-class families choosing to live in affluent areas that are distant from and socially disconnected from areas of poverty and social exclusion. Focusing upon young people living in socially deprived areas in northern England, MacDonald and Shildrick (2007) found that many young people – having left school and secured some form of work – moved from hanging out in the streets of their local area to spending time in bars and clubs in the town centre. However, for several young men in their study, the local area remained the focus of their leisure time into their early twenties. These young men, through repeated unemployment, 'lacked the sites through which to establish new, more socially varied or geographically spread, social networks' (MacDonald and Shildrick 2007: 344). As such, some young people are financially excluded from participating in various contemporary forms of leisure, which are often stereotypically associated with young people. This results in locally based friendship networks being the main focus of their lives as they tended to spend a lot of their time with other young men in the same situation as them. However, these engagements with 'street corner society' (MacDonald and Shildrick 2007: 346) often decreased with age and changed as young people transitioned into having their own families or to experiences of independent living.

MacDonald *et al.* (2005: 877) have also noted the 'localization of housing careers for young people in poor areas' with many showing the tendency to live close to extended family members within the neighbourhood or community in which they were brought up. Often, young people are embedded 'in close, locally-concentrated family and social networks' and appear to accept their experiences of social exclusion. Many struggled to gain or hold down employment which – when available – tended to be in the 'lower reaches of the service industry and routinized, factory production' (MacDonald *et al.* 2005: 881). Such work was a feature of family life as many fathers and mothers were employed in similar forms of work and so the youth labour market therefore should not be seen as separate. According to this research, 'unemployment, job insecurity and poor work have become common working-class experiences, rather than the preserve of a wholly excluded underclass positioned beneath them' (MacDonald *et al.* 2005: 882). Such experiences resulted in young people having to adopt informal strategies for finding work drawing upon their contacts and networks. Furthermore, although young people's working and leisure lives focused on their local neighbourhood, the unsteady landscapes of employment are increasingly influenced by the rapidly changing nature of regional and national economies, and often shaped increasingly by international economic forces. Overall then, there have been radical changes in the experiences and circumstances of working-class young people so much so that 'practices that once helped working class young adults "get on" now, at best, only help them to "get by" ' (MacDonald *et al.* 2005: 886).

Despite the increasingly challenging neighbourhood and community contexts in which socially deprived young people grow up, Hill *et al.* (2007: 17) argue that 'even in relation to poverty there are often redeeming features in the local environment'. They point to the role of 'social services, schools, recreational facilities, churches and community centres' as having 'important positive influences in optimising responses to poverty and intrafamilial difficulties' (Hill *et al.* 2007: 17). Furthermore, in research conducted by Shildrick (2006) in the north-east of England, neighbourhood was a key factor in how young people constructed social relations. Often this was related to the type of housing – council and private – and residential location, highlighting the important role that neighbourhood relations have for young people's senses of identity and resilience (Table 6.3).

Table 6.3 ***Resilience and young people***

Resources	*Protective mechanism*
Individual level	
Constitutional	Positive temperament Robust neurobiology
Sociability	Responsiveness to others Pro-social attitudes Attachment to others
Intelligence	Academic achievement and ability Planning and decision making
Communication skills	Developed language skills Advanced reading
Personal attributes	Tolerance of negative effect Self-efficacy Self-esteem Foundational sense of self Internal locus of control Sense of humour Hopefulness Strategies to deal with stress and anxiety Enduring set of values Balanced perspectives on experience Malleable and flexible Fortitude, conviction, tenacity and resolve
Family level	
Supportive families	Parental warmth, encouragement, assistance Cohesion and care within the family Close relationship with a caring adult Belief in the child Non-blaming Marital support Talent or hobby valued by others
Social environment level	
Socio-economic status	Material resources and facilities
School experiences	Supportive peer group Positive influence from teachers Academic or other successes
Supportive communities	Believes the individual's stress Non-punitive Provisions and resources to assist Belief in the values of a society

Adapted from Olsson *et al.* (2003: 5–6)

> Locality continues to act as a focus for some working-class cultural identifications, often amongst those who are in some sense marginal to production and to the collective solidarities generated elsewhere. Locality continues to act as a base for collective activity among working class adolescents, both in the sense of providing cultural identities . . . for many otherwise unnamed youth groupings – and constituting their 'social space' – 'the street', alleyways etc. which are public and less tightly regulated than other areas.
>
> (Clarke 1979: 251, cited in Shildrick 2006: 72)

Communities and faith

Senses of belonging to different faith communities often offer young people an alternative form of community. Christian Smith (2003) suggests that religion may have a positive influence on the lives of young people in the USA (Box 6.2). It provides them with moral order (moral directives, spiritual experiences and role models), learned competencies (community and leadership skills, coping skills and cultural capital) and social and organisational ties (social capital, network closure and extra-community links). In the USA, Smith *et al.* (2002) found that in 1995, of religious youth, 24 per cent were Catholic and 23 per cent Baptist, and in terms of attending religious services associated with their faith communities, 38 per cent of young people attended weekly

Box 6.2

Religiosity and young people in the USA

1. The majority of American youth are religious insofar as they affiliate with some religious group or tradition. Only 13 per cent stated in 1995 that they had no religion.
2. The number of American adolescents within the Christian tradition has been gradually declining over the last two and a half decades.
3. About half of American adolescents regularly participate in religious organisations in the form of religious service attendance and participation in religious youth groups.
4. On the other hand, about half of American youth are not religiously active.
5. Religious participation declines with age.
6. Adolescent girls tend to be somewhat more religiously active than boys.
7. Religious participation is somewhat differentiated by race.
8. Religious participation varies somewhat by region of residence.

Adapted from Smith *et al.* (2002: 609–10)

with 16 per cent attending once or twice a month, 31 per cent rarely and 15 per cent never at all. Also, young women tended to feature higher on all indicators of religiosity than young men, although there was a steady decline in religious participation with age (Smith *et al.* 2002: 605).

In the UK, Crockett and Voas (2006: 581) have noted that 'religious decline in twentieth-century Britain was overwhelmingly generational in nature; decade by decade, year by year, each birth cohort was less religious than the one before.' Although religious affiliation is declining, so too is regular attendance at places of worship and perceptions about the importance of religion in society. They observe that for the majority of young people, the likelihood of having an affiliation with a religion is connected with the affiliations of parents, with young people likely to follow their parents' denominations (with mothers appearing more influential than fathers). Research about religion more generally has highlighted that there has been a rise in alternative forms of spirituality in recent years (Woodhead and Heelas 2005) leading to the formation of different spiritual communities rather than those relating to traditional forms of religion. Mason *et al.* (2007) explored the spirituality of young Australians and found that although 49 per cent adopted a form of spirituality which they define as grounded in the tradition of a major world religion (43 per cent Christian and 6 per cent other world religion), 17 per cent affiliated with an alternative form of spirituality and 31 per cent affiliated with a humanist approach which emphasises the importance of human experience rather than focusing on traditional religion or other forms of alternative spirituality.

In the accounts of young British Muslim women, Dwyer (1999b) found that the primary construction of community adopted by the young women referred to the local 'Asian community'. The young women were positive about the local facilities, presence of other Asian families and how the community offered them safe spaces away from racism. However, the young Muslim women were also aware of the ways in which the local Asian community – beyond their immediate family – operated to monitor their everyday behaviours, particularly those associated with gendered norms. Furthermore, young Muslim men were also found to reinforce this by policing the personal conduct and behaviour of the young women. Overall, the community is simultaneously a source of strength and oppression. For some of the young women, the construction of a Muslim community 'offers possibilities for challenging parental prohibitions and challenging the boundaries of a narrowly confined local community through the languages of Islam' (Dwyer 1999b: 64).

Communities and crime

Young people tend to be imagined as criminal, out of control and to be feared, so there is a 'common association of youth with crime' (Pain 2001: 908). Young people are 'at once the most criminalised and the most victimised of all social groups' (Pain 2003: 161) and 'experience high rates of property crime, violent crime, harassment and fear' (Pain 2003: 151). Young people aged 16–24 are four times more likely to be burgled than people over 65. As is the case with children and older people, young people are often more at risk in private spaces. After all, home is more dangerous is terms of abuse and accidents than public space. Furthermore, young people 'not only face higher risks of victimisation but also greater socialisation into fear, having grown up in an era when crime has become a major reason behind parental controls on children's spatial and social experiences' (Pain 2001b: 909). Many homeless young people experience abuse whilst at home and whilst homeless (Pain and Francis 2004). However, there is often underreporting of young people's experiences of victimisation.

The idea that young people are involved in crime is not new. Indeed, the setting up of the juvenile justice system in the late nineteenth century was one response to this issue (see Table 6.4 for current

Table 6.4 ***Ages of criminal responsibility in Europe***

Belgium	18
Denmark	15
England and Wales	10
Finland	15
France	13
Germany	14
Italy	14
Luxembourg	18
Netherlands	12
Northern Ireland	10
Norway	15
Portugal	16
Scotland	8
Spain	14
Sweden	15

Adapted from Muncie (2004: 251)

variations in ages of criminal responsibility across Europe). Henderson *et al.* (2007) draw a number of parallels between the social positioning and value of young people in the early 1900s compared with today and in particular highlight that the dominant discourse about youth and crime focuses upon issues of poor parenting, lack of education and civil unrest. Henderson *et al.* (2007) drew samples of young people from five locations in the UK – an isolated rural area, a disadvantaged northern estate, Northern Ireland, a leafy suburb, and an inner city location in southern England. They found that in Northern Ireland and in the disadvantaged estate, 'the level of violence young people encountered (including domestic and street violence) was significantly higher than in other areas' (2007: 63). Attitudes also differed with young people in both of these locations seeing fighting as a normal aspect of their everyday lives. Young people recruited within the rural and suburban neighbourhoods were the least likely to mention experiences of violence or crime. 'The nature of violence and crime discussed across the sites included domestic violence, sectarian violence, gang violence, race-related and homophobic violence, bullying, and unprovoked street attack' (Henderson *et al.* 2007: 63). 'The findings indicate that violence occupies an important place in young people's moral landscape' (Henderson *et al.* 2007: 63).

For the young people living on the disadvantaged estate, discussions about shootings, muggings, robberies and violent assaults were a regular topic of conversation. Some young people avoided hanging about in the evenings due to fear of crime. Concerns about personal safety featured in many of these urban young people's accounts, as they also did in the discussions with rural young people. Furthermore, a form of violence regularly discussed by the young women, domestic violence, featured across all study locations but was most evident in Northern Ireland and in the deprived estate. Alongside identities associated with the urban–rural continuum and gender, 'racially motivated crime has increased in recent years as has homophobic crime, and incidence of domestic violence has also risen in the past 10 years' (Henderson *et al.* 2007: 66), demonstrating that young people's ethnic identities and sexual orientation also play a role in determining their experiences of crime and violence.

Public spaces are increasingly regulated and reinforced as adult spaces and 'young people are increasingly subject to control, regulation and surveillance on the streets, in the home *and* in public institutions; the very places where they are more at risk from harm but often unprotected' (Pain 2003: 165). As such, those places often associated with safety and security for young people are regularly some of the most dangerous places. Children are more likely to be the victims rather than the perpetrators of crime (Pain 2000, 2001b), resulting in the retreat from the public spaces of the neighbourhood and community. Related to this – and as Valentine (1997) demonstrates – although many parents feel the need to regulate their children's use of public space, many children feel that their parents are overprotective, over-emotional and naïve in their understandings of public space. For Pain (2001b), concerns about age and fear of crime require an intergenerational approach, thereby considering how stereotypes, assumptions and behaviours associated with children, youth and older people work together to influence patterns of fear.

'As a public issue and a daily experience, fear of crime profoundly affects the ways in which the spaces young people use are produced and regulated' (Pain 2003: 154). Yet, young people's fear of crime tends to be ignored as they are viewed as the problem. However, young people with particular social experiences and social identities are often more likely to experience victimisation compared with other young people. For example, young people who are identified as homeless or from ethnic minorities, those excluded from school and those deemed 'hard to reach' may have the highest levels of victimisation yet are often viewed as the most dangerous in terms of how they are represented in popular and policy contexts. In childhood, fear appears to have most influence over parental control, and can contribute to what Pain (2006) discusses in terms of popular discourses about 'paranoid parenting'. This means that 'by the time they are allowed autonomous mobility, adolescents have learned powerful lessons about safe places and spaces and safe times to be outside the home' (Pain *et al.* 2001: 909). Formative experiences of neighbourhood and community therefore have powerful influences over young people's lives and their inclination to be the victims or perpetrators of crime: 'the places and neighbourhoods they grow up in also play a part in the level of crime and violence they are exposed to and are potentially involved in' (Henderson *et al.* 2007: 71).

Percy-Smith and Matthews (2001) found that a regular source of concern, anxiety and fear for young people was associated with being bullied, particularly by older young people. In particular, the young people recount older young people 'barging in' on their games, displaying threatening or coercive behaviour, intimidation and name-calling. They also found that bullying occurred more frequently in inner city areas (57 per cent) compared with suburban locations

(42 per cent) and suggested that an explanation for this could include the lack of space available for young people in the inner city compared with the suburbs. It was in parks, around local shops and in their local streets that young people were most likely to suffer from bullying. Curtis *et al.* (2008) have also drawn upon existing studies to show that strong connections between crime and disorder can increase the likelihood of young people suffering mental health problems and psychiatric symptoms. In particular, they observe that living in a neighbourhood with a high crime rate, witnessing violence, perceiving the neighbourhood to be risky and direct experiences of victimisation can all contribute to mental ill-health.

Key themes

- Young people's experiences of neighbourhood and community are regularly marked by strict territorial divisions associated with identifications with specific places and identities.
- Young people who live in neighbourhoods where poverty dominates suffer from a lack of services, employment and opportunities yet often choose to live locally and are likely to spend much time within the public spaces of their neighbourhood.
- Although membership of most religious groups amongst young people is declining, communities of faith or spirituality are often important mechanisms for young people to experience a sense of community and identity.
- In contrast to dominant discourses about youth and crime, more young people are the victims rather than the perpetrators crime.

Project ideas

Map out the private, statutory and free provision of facilities within your local neighbourhood. What facilities are provided for people of different age groups and are any of the facilities specifically for young people?

What are the key territorial divisions in your local neighbourhood or community and explain the processes in operation which sustain such distinctions?

Why are neighbourhood and community important for young people's constructions of identity and how do these constructions relate to other scales of belonging?

What policies might help neighbourhoods and communities to be more inclusive to the needs and expectations of young people?

Suggested further reading

This is a brilliant book that explores the experiences and accounts of young people's transitions to adulthood in different neighbourhoods and communities:

Henderson, Sheila, Holland, Janet, McGrellis, Sheens, Sharpe, Sue and Thomson, Rachel (2007) *Inventing adulthoods: a biographical approach to youth transitions*. London: Sage.

This insightful report charts the multiple meanings and experiences of territoriality for young people growing up in Britain:

Kintrea, Keith, Bannister, Jon, Pickering, Jon, Reid, Maggie and Suzuki, Naofumi (2008) *Young people and territoriality in British cities*. York: Joseph Rowntree Foundation.

This article offers a rich insight into how young people growing up in Northern Ireland negotiate their use of space:

Leonard, Madeleine (2006) Teens and territory in contested spaces: negotiating sectarian influences in Northern Ireland. *Children's Geographies* 4(2) 225–38.

This is a rich and insightful exploration of young people's negotiation of growing up in poor neighbourhoods:

MacDonald, Robert, Shildrick, Tracy, Webster, Colin and Simpson, Donald (2005) Growing up in poor neighbourhoods: the significance of class and place in the extended transitions of 'socially excluded' young adults. *Sociology* 39(2) 873–91.

This is an excellent exploration of the intersections of age, race, gender and fear for people living in urban areas:

Pain, Rachel (2001) Gender, race, age and fear in the city. *Urban Studies* 5–6 899–913.

This is an interesting insight into young people's engagement with faith and religious belief:

Smith, Christian, Denton, Melinda Lundquist, Faris, Robert and Regnerus, Mark (2002) Mapping American adolescent religious participation. *Journal for the Social Scientific Study of Religion* 14(4) 597–612.

7 Nation

Rina: Last year my sister and my brother went to Mexico. I don't know, the people there expected them to be rich. It's like, they said, 'Oh, we don't have money.' So my sister had to pay for dinner, for everything. She came back broke. Everybody expected her to pay for everything because she came from here. (Hyams 2002: 464)

Faruk: Everything is Scottish about me; I mean what can I say. Yeah, I'm a practising Muslim, and I practice Islam, but that doesn't mean I'm not Scottish. I do all the things that other Scottish people do. I play football, I go out, I do this and that. There is nothing that I can say is not Scottish about me. (Hopkins 2007b: 68)

Sumita: I mean, I think the whole approach was wrong, going in and bombing people all through the night, this is just horrible, when you think of all the kids and it, I think it was wrong. I think they could have gone in a different way. I mean, the war in Iraq, how many people did they kill? Saying it was an accident, still people did die. (Hörschelmann 2008a: 145)

Youthful nations

The sentiments expressed by Rina, Faruk and Sumita show young people engaging with issues of nation, nationality and politics in different ways. Rina demonstrates how perceptions around being American or being Mexican influence how young Latinas' national identities are read in different ways and in different times. The views of Faruk highlight how he sees himself as belonging within a Scottish national identity whilst also practising Islam, and Sumita shows her engagement with political issues through articulating her opinion about the war in Iraq. For some, the sentiments of these young people might be surprising as children and young people are rarely heard discussing issues of nation, nationhood, nationality or politics. Contrary to common understandings about young people's lives which see them as distant and remote from politics and nation, this chapter explores the ways in which young people engage with issues of national identity, national politics and participation. Overall, the intention of this chapter is to demonstrate the ways in which nation works to structure young people's identities as well as the ways in which young people have an active role in influencing such processes.

It is important to look at how nation is defined. Benedict Anderson (1983) sees nations as 'imagined communities' of people who follow similar customs and speak the same language. Scourfield *et al.* (2006a: 11) observe that 'nations have an abstract quality: they cannot be grasped directly by their inhabitants and must instead be made meaningful through constantly circulating symbols and images.' Furthermore, Jones *et al.* (2004: 83) note that 'if nations are common communities of people that share certain cultural attributes and a particular territory, then we need to think of nationalism as an ideology that seeks to promote the existence of nations within the world.' Jones *et al.* (2004) suggest that nations often have a number of key characteristics. First, all nations have a particular geography inherent within their claims to particular territories, and this is often interconnected with themes such as place and landscape. Second, nations also share cultural qualities, such as shared histories that are used to invoke loyalty from people. Third, legal and economic processes play a key role in the creation and maintenance of nations. It is also important to distinguish nation from state; the latter used to refer to a geographical governed territory. Although many nations are also states, there are also many nations that are not states in themselves, such as Scotland, which is often referred to as a stateless nation.

An important theme of this work is the recognition that 'nation' is a social and political construction rather than something that is natural or given. As Jackson and Penrose (1993: 1) observe:

> Demonstrating that 'race' and nation mean quite different things in different places underlines the notion that both concepts are social constructions, the product of specific historical and geographical forces, rather than biologically given ideas whose meaning is dictated by nature.

Cox (2002: 167) notes that the idea that nations are constructed socially 'is a very odd, even bizarre, idea' as there is a very strong tendency to see 'nationality as something we are born with, something that is given to us.' The myth that nations are given is often reinforced by the ways in which people can become 'naturalised' members of another nation whereby a natural and biological phrase is adopted to define membership of national space. The recognition that nation is a social construction does not mean that 'nation' is a meaningless concept. Instead, as Penrose (1995) suggests, it might be helpful to focus on the processes associated with the formation of nations rather than taking for granted the category of nation itself.

Although the scale of nation may work to shape young people's identities, the ideology of nation is often experienced or mediated through particular places, such as memorials, museums or government buildings. Place is a common theme within understandings of nation. 'Place – whether thought of in generic or specific terms – is, therefore, critical to any understandings of nations and nationalism' (Jones *et al.* 2004: 92). Certain places often become associated with or are regarded as taking on the characteristics of the nation. These places may be particular locations, buildings or landscapes and much work has explored the importance of landscape, in terms of both meanings and values as well as the physical landscape as being important aspects of nation. There is also a tendency to associate rurality with the nation in that there is a sense in which the rural brings people closer to nation. Different nations have specific histories and geographies, many of which are bound up in myth, literature and sound.

Since nations are socially constructed, an important theme in academic research has focused upon the ways in which nations are maintained and reproduced. In this respect, nations have cultural qualities or attributes that are often used to promote nationality and senses of national identities amongst the population. Michael Billig (1995) developed the idea of 'banal nationalism' to refer to the often small-scale and mundane ways in which ideas of nation are reproduced on a daily basis. This may include the use of symbols or imagery on bank notes, cars or in other locations engaged with on a daily basis.

As Scourfield *et al.* (2006a: 9) note, 'expressions of national feeling' often rely on an 'assortment of cultural markers. These typically include nation-defining images, symbols, narratives, legends, sporting activities, famous personalities, proclamations of shared values or social solidarity, and so forth.'

In thinking about young people, place and identity in the context of nation, many of the institutions young people engage with – and schools in particular – are important agents in reproducing and maintaining the nation. It is important to recognise the role of the state in educating people, setting the context for the school day, structuring the national curriculum and promoting particular national languages. As Scourfield *et al.* (2006a: 1) observe, 'national feeling is not natural or instinctive in children but is consciously cultivated in them (by adults), hence, the assumption that a nation's schools are places where dominant discourses of national identity and history are promulgated.' Combined with this, most nations also have national public holidays, national memorials, national events and practices, and a sense of shared identity and commonality is often developed through national sporting events.

Youthful national identities

There is a very large body of scholarship – across the social sciences – which explores issues associated with nations and nationalism, and much of this work focuses on national identity. Scholars of nations and nationalism often emphasise the importance of recognising that there are many nationalisms. An important distinction is often made between civic and ethnic nationalism. 'Civic nationalism (sometimes called "territorial nationalism") is a modern, liberal phenomenon geared towards the creation and regulation of an efficient, social, economic and political unit' (Gruffudd 1999: 200). On the other hand, ethnic nationalism focuses upon people's sense of belonging and their ethnic characteristics such as their language, religion, traditions customs. As Gruffudd (1999) notes, feelings of identification with nation – whether civic or ethnic – remain a powerful form of identity for many people and this includes young people. Yet, just as young people are marginalised within studies about nations and nationalism, the same too can be said with regards to work about young people and national identities. Scourfield *et al.* (2006a, b) observe that there is little work about children's relations to nations in contrast to the volume of work about adults, and that a focus upon the nation is relatively marginalised within the sociology of childhood. Furthermore, Thompson (2001) contends that nations are too often viewed without attention being paid to the people inhabiting national space and so he has argued that people need to be 'put back into nations.'

Alongside considering the place of young people in discussions of national identities, it is also important to think about how we come to understand national identity and what having a particular national identity means for young people. A young person might claim to be Australian, American, British or Swedish by ticking a box on a questionnaire, but what does it actually mean to identify as Australian, American, British or Swedish? Are there certain qualities or characteristics associated with particular national identities? Anthias (2001) has suggested that researchers could usefully focus upon narratives of national identity in seeking to excavate deeper understandings of nationality. She proposes that researchers focus on narratives of location, dislocation and positionality and so problematise the status of identity, whilst also using identity as a meaningful concept. By focusing on young people's accounts of location, dislocation and positionality, they will be able to recount notions of belonging and so tell a story about the social categories that they use to locate themselves in particular places and times (Image 7.1). Three studies that have focused upon young people's narratives in the context of national identity are now explored; they are based on research in the USA, Wales and Scotland.

Working with young Latinas, US-born teenage girls of Mexican descent in Los Angeles, Hyams (2002) explored the young women's articulations of nation and national identity:

> With few exceptions, in the Latinas' stories about Mexico and interactions with Mexicans, the country and lives of its people are evaluated negatively and represented in homogeneous and oppositional terms to people and places on 'the other side', the US side of the border.
>
> (Hyams 2002: 464)

The young women regarded things associated with Mexico as devalued and as lacking, in contrast to the contemporary and abundant USA. This sense of superiority was also bolstered by family practices associated with sending remittances to family members

Image 7.1 Youthful national identities.

in Mexico and the help provided when visiting family in Mexico. The young women also often drew upon binaries associated with dirt vis-à-vis cleanliness in their comparisons between the USA and Mexico. There was therefore a strong sense amongst the majority of the young women that their American national identity gave them access to the 'land of opportunity' compared to Mexico which they characterised as lacking in such opportunities.

Hyams (2002: 467) also discusses the intergenerational interactions between the young Latinas and their parents, and in particular the ways in which their parents tell of their hardships 'back then' and 'over there'. This often led the young women to associate their parents' historical experiences with the current circumstances of young Mexicans, thereby rendering Mexican identity as fixed and unchanging. Furthermore, the young women also drew upon gendered and embodied comparisons believing Mexican young women to be interested only in getting married and having children, compared to themselves, who were more interested in earning money, buying a car, getting a good job and buying a house. As such, the young women saw themselves as forward looking and fashionable compared to their unfashionable contemporaries. This example of young women's national identity formation highlights the importance of identities being relational as well as the importance of the family in constructing and shaping young people's identities.

A second example of youthful national identities is Scourfield *et al.*'s research about national identity in Wales with children aged 8 to 11. They note that children 'encounter nation (and nationalist) discourse in school, at home, around their local communities and via the media they consume. Children are, therefore, immersed in environments where national distinctions are still very much on display' (Scourfield *et al.* 2006a: 64). The children – during a 'secret-ballot' card sorting exercise – regularly chose Welsh as a form of identity. They found that children in Wales drew upon the resources available to them in their discussions about Wales and Welshness. In particular, place of birth, language and sport were key resources for the young people.

Place of birth was the dominant marker of nationality for the children. They were aware that it was possible to 'feel' Welsh, for example, by living in Wales, even if not born there, and a few gave examples of themselves or their parents as having come to feel Welsh through living in Wales (Box 7.1). However, there was a general tendency for them to prioritise country of origin when referring to a technical nationality, either referencing their own birthplace or that of their parents. This apart, there was little in the way of distinctive cultural content for the category of 'being Welsh'. Rugby was frequently mentioned as a signifier of Welshness, yet nearly all the children said they supported both a Welsh national rugby team and an English premier football team – typically, Manchester United (Scourfield *et al.* 2006b: 586).

Scourfield *et al.* (2006a: 75) question what makes being Welsh distinctive other than language and location – 'apart from supporting a Welsh national sports team, being born within the country's borders or claiming an affinity with a language that has an almost unique connection to the borders to Wales, what else is there?' Unfortunately they do not provide a direct answer to this question, most probably because it is an exceptionally difficult – if not impossible – question to answer.

A third example of young people's national identity construction can be found in the experiences of young Muslim men from Scotland

Box 7.1

Youthful Welsh national identities

Place of birth was the dominant resource of national identity drawn upon by the children and they would often refer to their parents' place of birth rather than their own. Young people would therefore claim to be Welsh if both parents were born in Wales, and if they had a parent born elsewhere, they could often draw upon labels such as half Welsh and half English. These often simple choices were more complex for young people from minority ethnic backgrounds who would draw upon their religion, parents' country of origin and other forms of national identity, such as British, in talking about their nationality.

Alongside place of birth, young people also drew upon language and accent as resources for national identity. Many of the young people identified the ability to speak Welsh as being an important aspect of Welsh national identity although they argued that both Welsh and English speakers could be equally Welsh in terms of their national identity. The children from a region in north-west Wales where 80 per cent of the population speak Welsh tended to feel more distinctive than the rest of the UK compared with other young people involved in the study.

As Scourfield *et al.* note, sport is often an important resource for people's senses of national identity and is one of the few social activities where nationality comes to the fore. 'Sport was associated with patriotism and national triumph for the children' (2006a: 71). Many of the children recalled sporting events and occasions as important to their sense of Welshness. Some boys, however, tended to focus their sporting allegiances on the successful football teams of the English premier league given their interest in and engagement with football as their main form of sport. Some of the young people also commented – often in a perplexed way – on the ways in which some adults, and their fathers in particular, became very emotional with regard to sporting events, something which Scourfield *et al.* note that boys may feel more pressured to replicate than girls.

Scourfield *et al.* (2004, 2006a, b)

(Hopkins 2007b). In asserting a Scottish national identity, the young men used markers such as place of birth, length of residence, a commitment to place as well as upbringing and accent, and the vast majority of the young Muslim men involved in this research identified as Scottish Muslims. Furthermore, the young men mentioned drinking Irn-Bru (a popular Scottish soft drink), playing football and having an appreciation of Scotland's natural environment and people as factors that connected them to Scottishness (Hopkins 2004).

Alongside strong connections with Scottish national identity, the young Muslim men involved in this research mentioned aspects of their identities which made them less or not Scottish (Hopkins 2004). They mentioned having different ancestry, a different religion, not wearing kilts and not always participating in national events such as Burns Night. When asked if there were certain things that they would say are not Scottish about them, the most frequent response from the young Muslim men was to refer to the drinking, pub and club culture. There was a strong sense amongst the young men that the drinking culture in Scotland, itself a stereotype, played an important role in being Scottish and, because of their religious faith, they were unable to participate in such activities as so were excluded from engaging with particular aspects of their national identity. There were also concerns that the drinking culture promoted racism with people being more likely to act in a discriminatory way when under the influence of alcohol. This being said, a minority of young Muslim men mentioned participating in drinking and clubbing culture although they argued that this was not about attracting the opposite sex but was instead about having a good night out. A final factor which the young men felt uncomfortable about in terms of identifying completely with a Scottish national identity related to the lack of modesty that they see some Scottish people displaying in terms of behaviour and dress. Many of the young men emphasised dressing modestly as an important aspect of practising their religion, yet felt that what they saw as the sexually suggestive dress sense of some men and women in Scotland made them feel uncomfortable and so they distanced themselves from a strong association with Scottishness (Hopkins 2004).

National political engagement

Thinking about nation often leads to considerations relating to politics and international relations, political geographies or studies about political engagement with the nation. Just as there is little research about young people and national identities, political geographers have also tended to omit young people from their work. As Philo and Smith (2003: 103) explain:

> The sub-discipline of political geography has never shown any special interest in children and young people, for the understandable reason that people below voting age cannot and do not have much direct influence on the obviously 'political' phenomena and structures (to do with nations, states, federations, elections, geopolitics, boundaries and the like) that have long been the staple subject-matter of political geography.

Young people who are too young to vote cannot participate in the political process and so the assumption is that the system has little to do with them. This assumption is supported by the fact that young people who are of voting age are stereotyped as being disengaged and apathetic about politics. As Henn *et al.* (2002: 167) note, 'conventional wisdom suggests that young people are becoming increasingly disengaged from politics and the democratic system'. In some respects, such claims are true. For example, the number of young people voting in the 2001 UK general election fell more than for any other group (Park 2004) with an estimated 39 per cent of 18–24 year olds participating in the election (a fall of 27 per cent compared with the percentage of young people voting in 1997). Furthermore,

> [i]n comparison with older age cohorts, young people are less likely to vote in elections, less likely to be members of political organisations, express less interest in politics and are much less likely to offer a party political identification . . . Furthermore, young people are reported to be highly disillusioned with the operation of politics and display very low levels of system efficacy.
>
> (Henn *et al.* 2002: 170)

The conclusion here could be that young people are disengaged from participating in national politics and so are disconnected, apathetic and inert when it comes to engaging with nation. Some authors have attempted to explain why many young people choose not to participate in UK general elections. For example, Kimberlee (2002) developed four sets of explanations for young people's non-participation in general elections in the UK, identifying youth-focused, politics-focused, alternative value and generational explanations (Box 7.2). Fahmy (2003) found that young people tended to believe that politics was not accessible; regarded politics as irrelevant; felt that politicians cannot be trusted; and saw little point in political action. He explored approaches for increasing the participation of young people, including lowering the voting age, changing the structure of institutions in order to incorporate the participation of

Box 7.2

Explanations for young people's non-participation in UK general elections

Youth-focused explanations draw attention to young people's age or social background with the explanation pointing to their individual apathy or social class positioning leading to non-participation. This is the most popular interpretation employed in the mass media.

Politics-focused explanations highlight the barriers created by the state or political parties that exclude young people from full participation. Dated electoral systems and the failure of political parties to attract the interests of young people with politics and politicians being disengaged and irrelevant to young people's lives tend to be the focus of such explanations.

Alternative value explanations see young people being attracted to political issues that are outside formal politics. Furthermore, with increasing mobility and lack of direct access to local labour markets, this explanation focuses upon the loss of traditional local routes into political engagement. This is also a common theme in the popular media

Generational explanations point to changing social circumstances and transitions that restrict or prevent young people's participation. The explanation here is that young people are experiencing unique social circumstances – such as changing transitions to adulthood and changing family situations – which are working to reduce participation.

Adapted from Kimberlee (2002)

young people and increasing the number of opportunities people have to engage with formal political issues. The way in which decisions are made about the age at which young people can vote is a key consideration for youth policy (Vanderbeck 2008).

This focus upon explaining or remedying declining levels of national political participation can be contrasted with research that highlights the ways in which young people are engaged and interested in political issues. For example, White *et al.* (2000) worked with two groups of young people in the UK, aged 14–24, one group of first-time voters and another group who would be the next generation of first-time voters. In mapping young people's concerns, they found that there were four different levels: the personal, local, national and global. This clearly demonstrates that – amongst other factors – young people are concerned about and engaged in national political issues (Table 7.1). Recognising the ways in which young people are engaged with national

Table 7.1 ***Young people's concerns***

Global	Local
Relationships with other countries	Labour market and job opportunities
Wars and political unrest	Lack of social facilities
Conditions in other countries	Drug use and pushers
Nationalism and devolution	Crime and violence
Protection of the environment	Policing strategies and treatment
Preservation of wildlife	Family planning advice
	Unreliable transport provision
National	*Personal*
Education	School/college experience
Employment and work	Acquiring qualifications
Economy	Job availability and security
Drug laws and education	Lack of money
Crime and justice	Social life
Age limits	Discrimination
Housing availability and related issues	Drug and alcohol use
Environment	Personal safety
Animal cruelty	Relationships
Transport policy	Treatment of young people
Government and politics	Keeping fit and healthy
Media coverage	Self image/identity
	Under-age sex
	Invasion of privacy

Adapted from White *et al.* (2000: 7)

political issues provides two main challenges to the assumption that young people are apathetic about and disengaged from national politics. First, labelling young people as politically disengaged in this way relies upon looking at statistical data about voter turnout or membership numbers and so makes very broad assumptions about what non-participation actually means. Much research about political participation tends to rely on large-scale survey data and so makes assumptions and generalisations about young people's engagements with the spaces of the political (O'Toole *et al.* 2003). It may be that young people are politically engaged but are choosing not to participate for specific reasons. As Henn *et al.* (2002: 168) suggest 'some authors have concluded that young people are concerned about matters that are essentially "political" in nature, but that these concerns lie beyond the boundaries of how politics is conventionally understood'. Second, making assumption about young people's non-participation in voting or political party membership involves imposing an adult-centred understanding and approach to politics onto young people's experiences. The argument here is that young people may think of politics in different ways and engage with political issues in ways which are different from those traditionally understood by adults. As Henn *et al.* (2002: 169) observe, 'once young people are invited to discuss politics on their own terms, thus widening the definition of politics, then there is evidence of much higher levels of interest and activity by young people.' This is supported by O'Toole (2003) who identified three problems that limit understandings of why young people do not participate in politics (Box 7.3). Some clear themes emerged from O'Toole's (2003: 88) work with young people in the UK:

> The first was the importance that our respondents attached to consultation, with many feeling that consultation ought to be an integral part of the political process. Secondly, many felt that the social diversity they experienced in their local communities was not reflected by politicians and this was construed as indicative of an exclusive political culture. Finally, many respondents expressed an interest in, and attached importance to, local involvement.

Young people might be apathetic, but for Marsh *et al.* (2007) the crucial issue is that they are alienated from the current political system. For Marsh *et al.* politics is a lived experience that is structured by issues of class, gender, ethnicity and age: '[T]he economic, social and cultural resources that young people had access to shaped their experiences and these in turn shaped their definitions of politics and views of political institutions' (p. 212). It is not necessarily age but lifecourse events which influences what young people see as political – whether this be education, training, employment or community resources: there was a strong sense that politicians were not addressing the issues that are important to young people

Box 7.3

Three problems that limit fuller understandings of the complexities of youth non-participation

- Research about political engagement tends to operate within a narrowly defined conceptualisation of what is meant by 'politics'.
- Research tends to assume that non-participation is the result of apathy, yet people do not participate for a complex range of reasons and not only because they are apathetic.
- Research tends to overlook youth-specific explanations for non-participation.

Adapted from O'Toole (2003)

and that there was little opportunity for them to participate in the political process.

Although young people rarely talked about class, those with less economic capital were aware of their socio-economic position and saw this as a political issue. Indeed, Marsh *et al.* (2007) found that politics often entered deeply into the everyday lives of young people from less privileged backgrounds and these young people recognised that these influences were often very negative. In terms of gender, Marsh *et al.* (2007) observed that many working-class young women were very aware of their social positioning, whilst other young women and most young men argued that inequality between the sexes was decreasing. Respondents also found issues of race and ethnicity to be political and again this was found in their everyday lives through ethnically segregated neighbourhoods, white political institutions and political responses to racism and discrimination:

> Our respondents were cynical about politics and politicians. There was a dominant perception that politicians were disconnected from the lives of ordinary, especially young, people; this was exacerbated for most by the inadequate number of black and minority ethnic, female or young politicians.
>
> (Marsh *et al.* 2007: 216)

Another set of examples of young people's political agency is evidenced through their protests against the second Iraq war in 2003. As Such *et al.* (2005: 301) note, 'ostensibly the protests were organised by young people and, shockingly for some adults, took place during school time.' In many cities in the UK, Germany and across the world, young people organised protests and Such *et al.* (2005) conducted a content analysis of a section of the British press between February and May 2003 in order to explore the representations of young people's protesting against the war. A key finding of their work was the discomfort that many adults felt with regards to young people protesting against the war. This was due to a number of factors including adult concerns about maintaining their dominance over children, young people missing school due to participating in protests and adults seeing children as lacking political agency. Young people's engagements with these 'global issues' are discussed further in the next chapter.

Focusing upon the experiences of young deaf people, Skelton and Valentine (2003) have highlighted the multiple ways in which their everyday lives engage with political issues. They found that a number of the young people involved in their research were engaged in volunteering and campaigning activities through assisting at local clubs or sports groups. Through this engagement, young people were offering support whilst also bringing about change. Some young people actively used British Sign Language (BSL) in public as a mechanism for demonstrating their deaf identity and therefore were engaging with their everyday experiences in political ways. Those young people who experienced discrimination often actively resisted such treatment and also participated in campaigning for the official recognition of BSL.

Drawing upon the Belgian Youth Survey, Quintelier (2008: 355) argues that 'membership of voluntary organizations makes people more likely to participate in politics in multiple ways'. It is suggested that by participating in voluntary organisations, young people acquire skills that are required for political participation such as political decision-making and politically relevant attitudes. The sense is that these organisations and agencies socialise young people into political participation. The type of voluntary organisation also matters in terms of the ways in which political participation might be promoted. Young people in school groups, interest groups or cultural and religious groups often share a sense of obligation with issues external to the group, whereas sports or music groups often have a social focus, are easy to leave and less likely to develop the skills necessary for political participation. Quintelier (2008) also discusses the significance of young people having a leadership role within an organisation as having a key role in their tendency to participate politically given their experiences of organising activities, facilitating events and gathering and collating information. Furthermore, young people who are members of more than one organisation are more likely to participate politically.

Beyond nation

The discussion so far has focused upon the ways in which young people engage with nation. However, there is also research that explores the ways in which people's identifications are moving beyond nation to focus upon particular regional or sub-regional affiliations or supranational and multinational forms of identification. As such, it is important to recognise that nation is contested and is often subject to inner disagreement, regional tension as well as multiple connections with other nations. So, as well as connecting with nation, young people identify with other scales of belonging such as those formed through devolution, regional nationalisms or supranational connections. The examples of youthful Welsh and Scottish national identities point to the importance of devolution for the ways in which young people construct their identities and the growth of European Union has also offered a new form of territorial identification.

The Runnymede Trust (1998) found that amongst young people in the UK, 60 per cent thought of British people as European although 61 per cent said they would never or rarely identify themselves as European. Young people identified freedom of movement across borders and free trade in Europe as a positive as well as mentioning strengthened military support and economic prosperity. Some also drew upon references to European sport. At the same time, however, there were concerns about the sovereignty of Britain and many young people raised the ban on British beef as a concern. Overall, the single currency and easier travel were the most discussed issues, and European law was generally seen in a negative light.

Grundy and Jamieson (2007) have explored young adults' understandings of European identity, drawing upon a sample of 18–24-year-old young people from Edinburgh, Scotland. The random sample of young people was compared with young people of the same age who were involved in courses in European languages, European law or European studies. Overall, although both samples of young people identified very weakly with European identities, they felt stronger about being European or their attachment to Europe compared to European citizenship. Those without a strong European identity were reasonably positive about the European Union. Those who explicitly identified with Europe – in particular one young person from the random sample and seven from the second sample – demonstrated that they were what Grundy and Jamieson (2007: 671) call 'passionate Europeans': 'they could cite situations in which they felt particularly conscious of being European, had well-developed views about a range of European issues and spoke, often with feeling, about their various visions of the future of European'.

These young people recognised that they were different from most of their peers. The sources of their interests in Europe included early childhood experiences, interests in languages or travelling as a young adult – 'recurrent themes involved friendships, shared experiences, emotions and meaningful communication with European nationals from other countries' (Grundy and Jamieson 2007: 677). That being said, for the majority of the research participants, they saw themselves as residents of Edinburgh and as Scottish nationals rather than being European. Many were unsure about what Europe meant beyond its geographical boundaries. Alongside discussing their connections with Europe, many of the young people observed that they often had to correct problematic assumptions made about Scotland. For example, they discussed challenging people for assuming that they were English, or that Scotland is in England or that England and Britain are the same thing.

Alongside supranational connections, there are strong arguments that young people's active participation and engagement with political issues should be grounded in local issues and contexts. Matthews and Limb (1998) have explored the formation and nature of youth councils and youth forums in the UK where a number of national organisations played influential roles in the development and promotion of these councils and forums, most of which operated in relatively local or regional contexts. They found different histories and contexts for youth councils and forums within the different national contexts of the UK as well as different types of youth councils and forums given the lack of a national model. In particular, they identified six types: feeder or constituent organisations; shadow organisations; issue-specific organisations; community-development organisations; group-specific organisations; and young people-initiated organisations.

Matthews and Limb (2003) also explored the views of 18-year-old young people about youth councils through the use of interviews and focus groups. In doing so, they drew upon the four core principles of political participation which are informed by Article 12 of the United Nations Convention on the Rights of the Child (UNCRC). In terms of making a 'visible commitment to . . . involve children and young people', coupled with the provision of adequate resources, they found that even when appropriate resources are in place, the key challenge was ensuring that the adults working with children were sufficiently skilled to enable and facilitate young people's active involvement and participation. The importance of time, resources and commitment are crucial. They also raised concerns about the lack of representativeness of the young people involved in youth councils and were concerned that these become the only mechanisms for accessing young people's views. Some youth councils were set up without making it clear why they were established and what young people's involvement would entail.

A strong argument within work about young people's participation and engagement is that it should be bottom-up and youth-led rather than being controlled and structured by adults. As Hart (1997: 4) notes:

> A nation is democratic to the extent that its citizens are involved, particularly at the community level. The confidence and competence to be involved must be gradually acquired through practice. It is for this reason that there should be gradually increasing opportunities for children to participate in any aspiring democracy, and particularly in those nations already convinced that they are democratic. With the growth of children's rights we are beginning to see an increasing recognition of children's abilities to speak for themselves. Regrettably, while children's and youths' participation does occur in different degrees around the world, it is often exploitative or frivolous.

These observations about young people's participation led Hart (1997) to develop the ladder of participation (see Figure 7.1) to identify the ways in which young people might participate ranging from manipulation and decoration or tokenism (i.e. non-participation) to varying degrees of participation ranging from their involvement being assigned to youth-initiated decision making being shared with adults.

Matthews *et al.* (1999a: 136) observe that there is 'intransigence in some quarters' about whether children's rights to participate is appropriate. Some adults are reluctant for children to participate in decision making as they worry about how this will threaten the presumed authority and power of parents within the family setting. Others feel that giving children too much responsibility takes something away from their childhood, which they see as being characterised by a freedom from responsibilities. Others believe that children should not have rights until they can deal with responsibilities. Another argument is that 'children are incapable of reasonable and rational decision-making, and may display incompetence confounded by their lack of experience and a likelihood that they will make mistakes' (Matthews *et al.* 1999a: 137). Yet, Article 12 of the UNCRC

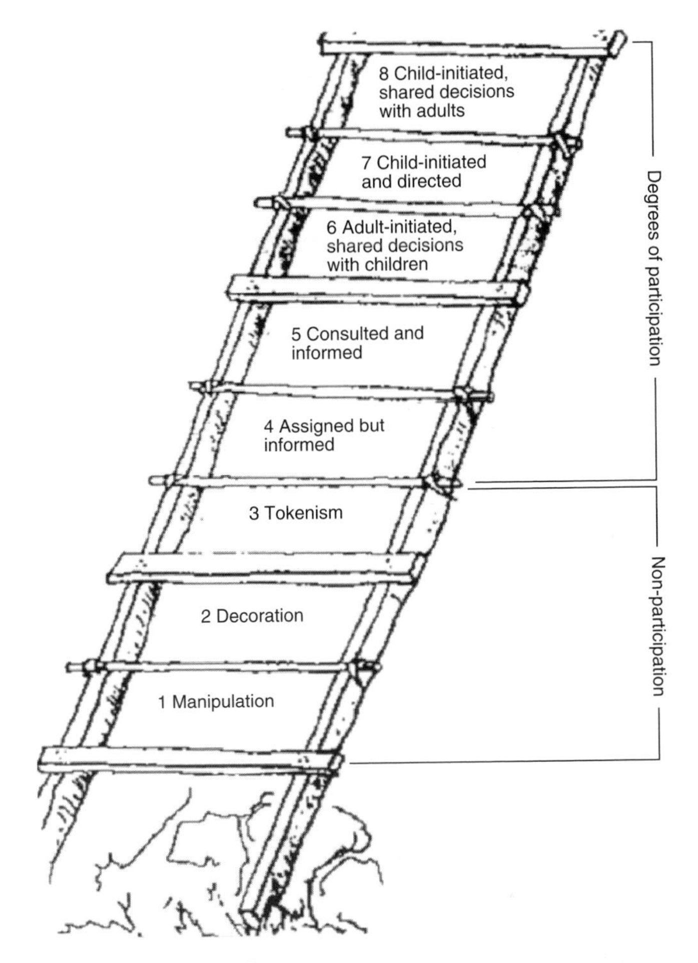

Figure 7.1 Hart's ladder of participation.

Source: Hart 1997

emphasises the importance of children's rights to have their views taken into consideration with regards to any issues or procedures relating to them.

Key themes

- Nations are not natural or given entities and are instead socially and politically constructed. It is important to explore the processes which give rise to nations and different forms of national identity rather than focusing solely on the category of 'nation' itself.

There is a need for research to include children and young people in understandings of nation, nationality and nationhood.

- Nations influence young people's lives in many ways such as through the structure of the national curriculum, national holidays or through stories or myths about the history of development of the nation.
- A deeper appreciation of national identity amongst young people may be developed by focusing upon young people's narratives and talk or through participatory research rather than through the employment of surveys. Young people's senses of national identity are often informed by subtle and everyday aspects of their lives (banal nationalism), such as where they were born, their parent's background, the languages they speak and the sporting teams they support.
- Young people's senses of national identity and belonging are often shaped by gendered assumptions and practices, senses of having (or not) shared religious practices with other young people or through attitudes and values present within society with which they do or do not agree with.
- Young people are often assumed to be politically disengaged, apathetic and inert due to their tendency not to vote at elections. However, such an argument has been critiqued for placing an adult understanding of politics onto young people. Non-participation in national politics does not necessarily mean that young people are apathetic. Adopting a broader understanding of politics and the political opens up the possibility of seeing young people as political agents who engage with politics in different and creative ways.
- Young people also identify with and participate and engage in ways that go beyond the nation, including supranational forms of identification or engagement that is local or regional in character. They can often feel marginalised and disenfranchised when participation is adult-centred and when they are not consulted about what their participation is for and what it means. Young people's active participation is often valued when it adopts a youth-centred, bottom-up approach.

Project ideas

What are some of the ethical issues involved in doing research with young people about national identities or national political engagement?

What do you feel your national identity is? How important do you think your national identity is compared with your other identities? Why do you subscribe to this form of national identity? How would you explain it to an outsider and what markers of national identity would you use in such an explanation?

What are the different explanations for young people's declining levels of participation in national politics? Is there an explanation that is more satisfactory than others, and if so, why?

What are some of the political and economic changes that have resulted in forms of identification which are beyond the nation?

What future changes in social, political and economic relations do you think will shape how young people engage with nation and what role will young people play in such changes?

Suggested further reading

This article explores the ways in which Muslim youth are simultaneously included within and excluded from Scotland and Scottishness:

Hopkins, Peter (2004) Young Muslim men in Scotland: inclusions and exclusions. *Children's Geographies* 2(2) 252–72.

This article discusses the ways in which young Latinas construct senses of national belonging through difference and place:

Hyams, Melissa (2002) 'Over there' and 'back then': an odyssey in national subjectivity. *Environment and Planning D: Society and Space* 20 459–76.

This book provides a useful introduction to key themes in political geography including understandings of nation and nationhood:

Jones, Martin, Jones, Rhys and Woods, Michael (2004) *An introduction to political geography: space, place and politics*. London: Routledge.

This excellent book provides a rich insight into the relationships between youth and politics in the UK:

Marsh, David, O'Toole, Thérèse and Jones, Su (2007) *Young people and politics in the UK: apathy or alienation?* New York: Palgrave Macmillan.

This article explores the multiples set of relationships between young people, politics and the political:

O'Toole, Thérèse (2003) Engaging with young people's conceptions of the political. *Children's Geographies* 1(1) 71–90.

This useful book explores how children in Wales understand their national and local identities:

Scourfield, Jonathan, Dicks, Bella, Drakeford, Mark and Davies, Andrew (2006) *Children, place and identity: nation and locality in middle childhood.* London: Routledge.

8 Global

Tariq: We were victims of what happened. We had loved ones stuck in the World Trade Center. We couldn't get contact to our loved ones. And on top of that, we were accused. So we were double traumatised. (Peek 2003a: 282)

Nina: . . . they are all very rich and drive expensive cars, are all well dressed and . . . I mean, it's all nicely furnished

Katja: But that's television

Nina: Yes, and you don't know what they are really like. (Hörschelmann and Schafer 2007: 1863)

Global youth

In the quotes above, Tariq talks about his views of the events of 11 September 2001 as a Muslim studying at a university in New York whilst Nina and Katja discuss their views on the ways in which lifestyles are portrayed in US soaps that they watch in their spare time in Leipzig in East Germany. The views expressed by these young people point to the complex ways in which their everyday lives are influenced by global issues, events and forces. This is hardly surprising given that there is a very strong sense that our everyday lives – and those of young people – are strongly influenced and shaped by global issues (Miles 2000). Commonsense perceptions about the influence of globalisation tend to assume that the global is all-powerful and that other transnational forces are key influences on how people live their lives (Miles 2000). As Short and Kim (1999: 75–6) suggest:

> the fact that people across the globe are watching CNN and MTV, that McDonald's franchises are opening around the world, and that many Hollywood films dominate the world film market, are taken as indisputable evidence of the Americanization of the world.

Furthermore, Ritzer (1993: 1) describes McDonaldisation as 'the process by which the principles of the fast-food restaurant are coming to dominate more and more sectors of American society as well as the rest of the world'. The sense here is that social, cultural and economic homogenisation is occurring as young people in different parts of the world come to experience various aspects of their lives in increasingly similar ways. Another common viewpoint 'suggests that globalisation in the cultural sphere produces Western hegemonic structures at the expense of local cultures. Cultural globalisation has been characterised as the flow of ideas, products and practices from the Western "core" to the "periphery" of non-Western locations' (Kehily and Nayak 2008: 328).

Contrary to these understandings of the influence of globalisation, a large volume of literature has emerged, offering more complex readings of the role and influence of globalisation (e.g. Appadurai 1996, Katz 2004). As Appadurai (1996: 42) suggests:

> The globalisation of culture is not the same as its homogenization, but globalisation involves the use of a variety of instruments of homogenization (armaments, advertising techniques, language hegemonics, and clothing styles) that are absorbed into local politics and cultural economies, only to be repatriated at heterogeneous dialogues.

His argument proposes that global forces 'tend to become indigenised in one or another way' (Appadurai 1996: 32) at the same time as their influence is spread across the world. Globalisation is therefore not a one-way process and it may be encouraging the homogenisation of cultures and societies. It also encourages heterogenisation. 'As the world becomes more the same, difference is prized, visited, created, commodified' (Short and Kim 1999: 80). Short and Kim (1999: 78) identify the process of the reterritorialisation of culture as 'a series of processes ranging from diffusion from their origin across borders (spatial, temporal and cultural) to establishment in a new place in a new form'. Overall, then, McDonalds in Moscow is not simply the Americanisation of Russia, but is a new form of consumption and identity that has taken root in a new place. McDonalds in Moscow will probably be used, experienced and thought of in a very different way from McDonalds in Paris, New York or Sydney.

Clearly, globalisation is multifaceted and complex, including diverse processes that shape finance, culture, communication technology and social relations. These diverse processes are organised into 'global cultural flows' by Appadurai (1996: 33) who identifies five fluid and irregular 'scapes': ethnoscapes, mediascapes, technoscapes, financescapes and ideoscapes (see Box 8.1). These scapes relate to various aspects of migration, technology, finance, media and ideologies, each of which is configured in complex ways to shape and influence how people experience their everyday lives.

Alongside the hype associated with globalisation, there is a critical body of scholarship that speaks directly to the place of young people in such debates (Aitken 2001b, Katz 2004, Wyness 2006). At the same

Box 8.1

Five dimensions of global cultural flows

Ethnoscapes are about the 'landscape of persons who constitute the shifting world in which we live: tourists, immigrants, refugees, exiles, guest workers, and other mobile groups and individuals constitute an essential feature of the world and appear to affect the politics of (and between) nations to a hitherto unprecedented degree'.

Technoscapes are the 'global configurations, also ever fluid, of technology and the fact that technology, both high and low, both mechanical and informational, now moves at high speeds across various kinds of previously impervious boundaries'.

Financescapes relate to the ways in which 'the disposition of global capital is now a more mysterious, rapid, and difficult landscape to follow than ever before'.

Mediascapes refer to newspapers, magazines, television programmes, and associated forms of media which are increasingly available throughout the world and which often result in particular images of the world being created.

Ideoscapes refer to different ideologies operating globally and most often are those connected with states or movements associated with power.

Adapted from Appadurai (1996: 33–4)

time, however, I was very surprised when writing this chapter to find that most books about young people pay no attention to issues relating to globalisation (see Aitken 2001, Katz 2004 and Wyness 2006 for rare exceptions). This is probably partly due to the fact that work about globalisation tends to focus on economic and political issues coupled with the assumption that young people play little or no part in large-scale, transnational processes. Pain and Smith (2008) have observed – with regard to work about fear – the tendency for academic scholarship to focus either on issues connected with international relations or on questions about embodied, local and grounded experiences. They argue that there is a need to bring these approaches together in order to advance understandings of fear. The same too could be said about young people, place and identity in the context of the global as there is a clear need to integrate young people's perspectives and experiences with broader discussions about globalisation and the influence of global processes. As Cahill and Katz (2008: 2809) note: 'writing the everyday experiences of young people into global analyses foregrounds an embodied and situated geopolitics.' Three key themes to consider are the ways in which young people's lives connect with globalisation, the nature and pace of change and young people's engagement with the mass media.

In exploring young people's negotiations of globalisation, it is clear that their lives increasingly involve some form of connection with global issues, more so than in any previous generation. As Wyness (2006: 62) observes:

> In examining globalisation from a Western vantage point, much has been made of the way that growing up has become a process incorporating a much wider range of global reference points. One of the most significant reference points is the mass media, through which children and young people become a more integral part of a globalised consumer culture.

Arguably one of the most positive influences of an increasingly global focus has been the ratification of the United Nations Convention on the Rights of the Child in 1998 which fully clarified children's rights to participate in society (Matthews *et al.* 1999a). This international convention comprises 54 Articles detailing the rights of children, including their right to express an opinion and to have that opinion taken into account in any matter or procedure affecting them (Article 12), the right to freedom of expression (Article 13), the right to association and assembly (Article 15), the right to appropriate information (Article 17) and the right to an education (Article 28) which will encourage responsible citizenship.

Key themes within discussions about globalisation focus upon the nature and pace of change coupled with the stark inequalities that such processes often create. As Jeffrey and Dyson (2008:1) note:

> Global social and economic change is rapidly altering people's experience of youth. Widespread unemployment, new health risks, and political conflict are reshaping the social landscape in which children and young people grow up. At the same time, governments, nongovernmental organisations (NGOs), and the media are centrally concerned with disciplining youth, for example, through the circulation of negative images of young people.

The speed of change and negative representations of youth both contribute to increasing inequalities between young people living in different places which often relate to who young people are (their identities and socio-economic position) and where they are. As Ruddick (2003: 356–7) observes, '"Globalization" presumes, in some places, to tie our lives more closely together, through more elaborate, intensive networks of trade and exchange and, in others, to separate us through more intransigent means of exclusion.' Such inequalities and exclusions are often associated with the rise of neoliberalism which Wills (2005: 577) defines as:

> an economic doctrine that favours free markets, the deregulation of national economies, decentralization and the privatisation of previously state-owned enterprises (e.g. education, health). A doctrine that, in practice, favours the interests of the powerful (TNCs) against the less powerful within societies.

A key question to ask here then is whether young people are included or excluded, and why.

One of the primary ways that globalisation has influenced young people – as identified by Wyness (2006) – is the global mass media reaching into young people's lives through a variety of formats such as television programmes and the internet. In this sense, young people are closer to consumer culture and are seen as the 'quintessential consumers' (Wyness 2006: 63) who are in tune with up-to-date technologies and engage with global media. The ways in which children are encouraged – through television and video advertising – to participate in markets through consumption has led Aitken (2001b: 123) to identify the 'unchildlike child' to refer to the disruption in the boundaries between childhood and adulthood and an associated moral crisis about the increasing tendency of children to behave like adults.

In this chapter, I initially focus upon global events and global lives in order to explore the ways in which global forces and processes shape and are shaped by young people as they negotiate their multiple identities in different places and times. I then focus upon global–local connections in order to explore the ways in which young people negotiate the global locally. Overall, the central argument of this chapter is that young people's experiences of global forces, processes and events are local. Although young people are living in and experiencing a global world during their everyday lives, their experiences of globalisation are shaped by their local lives and by the ways in which they take up, shape and resist global forces.

Global events

The events of 11 September 2001 were clearly felt most directly by those living or working close to the World Trade Center or those who had friends or families caught up in the events. Drawing upon focus groups and interviews, Peek (2003a) explored the reactions and responses on university campuses in New York to these events, focusing specifically on the experiences of Muslim university students. Although the students recorded satisfaction with regard to the ways in which their universities responded to the events through offering support and promoting a positive campus environment, nearly all students did not feel safe in New York compared with how they felt on campus. The concerns of the students focused on the use of public transportation and worry about travelling alone, which some students responded to by relying on friends as travel companions in order to promote their sense of personal safety and social well-being. The students' concerns were often heightened through stares, nasty glances, verbal harassment, or, in two cases, physical assault. Although there were negative experiences, students also recalled neighbours or even complete strangers reaching out to them to offer support.

Peek (2003a) also explored the ways in which intergenerational relations were influenced by the events with many students mentioning that their families were very concerned for their safety and suggested taking time off before returning to study. The most frequently recorded concern of the young people with regard to their families focused on their parents' requests that they change their physical appearance in order to reduce the likelihood of them being identified as Muslims. Men reported that family members suggested that they shave their beards and young women were advised not to wear a headscarf. However, the topic that generated most heated discussion amongst the students was the media. Many felt that their religion was persistently misrepresented, generating misunderstanding and hostility between communities:

> Indeed, the media shapes public opinion and thus frames social reality, and because so many Americans were unfamiliar with Muslims and the Islamic faith, they turned to their usual media sources for a better understanding.
>
> (Peek 2003a: 280)

Initially, most students thought that the events of 11 September 2001 happened by accident. This soon changed to sadness, fear and anxiety, particularly when these students heard that Muslims were involved in what happened. Peek (2003b) notes that some of students turned to God after the attacks whilst also participating in campaigning projects and events to openly display their opposition to what had happened. This resulted in several participants feeling an increased sense of community as Muslim students (Peek 2003b), although this was tempered by the ways in which they felt increasingly scrutinised during their everyday lives. Students also recalled feeling excluded from mourning and grieving as they felt marginalised from the emotional responses to the attacks as a result of their religious identities (Peek 2003b).

Although the attacks on the World Trade Center on 11 September undoubtedly had an impact on young Muslims living in New York, the response of different social groups to the events has also influenced other groups of people in New York and the USA as well as communities in other parts of the world. Many young people lost relatives, family members and friends in the attacks and the global nature of the events has shaped the lives of young people from a range of places. In particular, young people identified as belonging to particular religious and ethnic minority groups have been the targets of racism, discrimination and harassment. Drawing upon participants' observations and individual and group discussions with young Sikhs in the upper Midwest of the USA, Verma (2006) found that Sikh youth were the subject of increasing levels of racism and harassment following the events of 11 September 2001, as the perpetrators assumed that they were Muslims rather than Sikhs. Sikh youth were found to be concealing their religious identities in attempts to avoid becoming the targets of racist attacks. As Verma (2006: 100) observes, 'youth become the unseen victims of disquieting global trends.'

Young people in religious and ethnic minority communities outside the USA were also the victims of racist abuse and intimidation following the global events of 11 September 2001. Two examples of this are evident from research that has taken place in Glasgow and Edinburgh, Scotland, and in Sydney, Australia. Drawing upon research in Scotland with young Muslim men, it is clear that the young men felt that markers of Muslim identity had heightened in significance following the events of 11 September 2001. In particular, the young men felt that their beards, skin colour and dress choices were aspects of their embodied identities which are powerful influences in determining their experiences (Hopkins 2004). Like those involved in the research by Peek (2003a, b), there was a strong sense that those who openly displayed markers of their religious identity would be more likely to become the victims of racist attacks than those who were less explicit in displaying their religiosity (Hopkins 2004).

Similarly, in Sydney, Australia, Noble (2005) discusses the social incivilities experienced by Australians of Arabic-speaking and Muslim backgrounds since the events of 11 September 2001. These incivilities included behaviour that respondents found rude or insulting, such as 'name-calling, something said aggressively, sometimes not, jokes in bad taste, bad manners, provocative and offensive gestures or even just a sense of social distance of unfriendliness or an excessive focus on someone's ethnicity' (Noble 2005: 110). In particular, respondents felt discomfort as a result of these experiences, yet feeling comfortable is an important aspect of young people's identities, sense of well-being and influences the extent to which people feel at home in the world:

> [I]f social discomfort reflects our capacity to make ourselves at home in the world, and the power of others to shape that capacity, then this has implications for understanding forms of national and social belonging.
>
> (Noble 2005: 117)

An important theme in this work then is the way in which young people's identities are interpreted by others, how young people actively present their different identities and choose to articulate these in different ways, in different times and in response to global events such as 11 September 2001. Pain (2009) argues that it is important to consider how fear influences mobility, behaviour and lifestyle, alongside its impact on people's emotional and social well-being. Furthermore, it is crucial to consider the local and grounded influence of global events and thereby appreciate how geopolitical events are responded to, felt and experienced in emotional ways; this is what Pain (2009) calls 'emotional geopolitics'.

Although particular groups of young people may suffer discrimination and intimidation as a result of the ways in which their various identities are interpreted, it is also important to recognise that young people are not just passive victims of global events and process. Instead, young people often respond openly and creatively by resisting, contesting and challenging global issues and events (Hopkins 2007a, Hörschelmann 2008a, b). Connecting with work in the previous chapter about the ways in which young people are often competent and knowledgeable about political issues, there is much evidence of the ways in which young people respond to global events in political ways, therefore actively engaging with geopolitics. The young Muslim men who participated in focus groups and interviews in Glasgow and Edinburgh in Scotland often discussed their responses to the events of 11 September 2001. In doing so, they displayed their knowledge and understanding of geopolitical issues as well as articulating their suspicion about the strong relationship between George Bush who was then President of the USA and Tony Blair who was Prime Minister of the UK at the time. They also sought to distance themselves from the global Muslim umma (community) in an attempt to distance themselves from those involved in the attacks (Hopkins 2007a). It is clear that

> the young men take up a range of strategies in order to resist the influence of racist politicians motivated by the events and aftermath of 11 September 2001 and the subsequent circulation of racist and Islamophobic literatures and images. These strategies range from tactical voting to asserting the peacefulness of their religious faith, from protesting in marches to writing letters and sending petitions to politicians.
>
> (Hopkins 2007a: 1130)

A further example of young people's response to global events is evidenced in Hörschelmann's (2008b) work with young East Germans in the city of Leipzig (see Box 8.2). She worked with 15 groups of young people recruited through youth clubs in different parts of the city with the majority of the participants aged between 12 and 18, and argues for greater recognition to be given to young people's political engagements. This work demonstrates that young people resisted the war in Iraq through a number of key sites that worked to bring global and distant issues closer to the centre of their lives. In particular, Hörschelmann (2008b: 594) draws attention to the media, school, the city, the home and the body as sites where young people 'encountered, negotiated, embraced or challenged the policies of international state actors in their everyday lives.'

To recap, young people may be the victims of global events as they are personally attacked or discriminated against according to who they are and where they are. However, as I have argued, young people are not simply victims of such events and often respond in politically engaged and ethical ways as evidenced in the examples discussed above. Furthermore, although young people are often concerned about their own well-being and security, research has demonstrated that they show more concern for those who suffer more directly the consequences of global events. Hörschelmann's (2008a) work with students in Bradford in the UK demonstrates that, for many, a primary concern was not their own safety but the outcomes of the Iraq war for those directly involved. Family connections for some, and a sense of injustice for others, motivated these concerns.

Global lives

An important theme in the small amount of work about young people that draws attention to globalisation is the way in which young people negotiate global issues and global forces in their everyday lives. This work often focuses upon how young people construct and contest their identities as well as how their everyday lives take place in the context of globalising forces and global issues. As such, it focuses on the ways in which global forces are negotiated in everyday local lives (e.g. Nayak 2003a). Important work by Hörschelmann and Schafer (2005, 2007) has explored the ways in which young east Germans negotiate global processes in the context of the transition from socialism to capitalism. Working with 101 young people aged between

Box 8.2

Young East Germans engaging with the global locally through critical geopolitics

Media – television media as well as newspapers and the radio were the main sources of information for young people about the Iraq war. Young people's levels of consumption of news media increased as the invasion of Iraq took place and the media offered the young people the language and frameworks that they used in their discussions. The young people often discussed the political personalities involved in the events that unfolded and felt increasingly disempowered and excluded from the power and control of the USA, a feeling that was exacerbated by what many felt was a very unfair series of events. At the same time, however, some young people were bored by the constant news coverage of the war and many talked about agreeing with their parents' perspectives on the situation.

The city – ten of the young people who participated in the research took part in organised protests in Leipzig, often as part of larger friendship groups. For some of the young people, these protests were sites of action where they could actively resist international politics. These actions placed the young people in protests against the Iraq war as well as being a part of a history of collective action in their own country.

The home – the home is the main place for media consumption for many young people as well as being a location where they interact with their friends, parents and siblings. The photographs taken by young people often showed their own bedrooms and some contained anti-war images such as Stop the War posters

Alongside these key sites, the discussions of the young people involved in the research also demonstrate the role of age and gender in influencing how the consequences of war are perceived. Young people often objected to the ways in which innocent children were the victims of ill-informed adult decisions. Although at once distant, the war was also close by for some young people who had friends serving in Afghanistan or for young people who felt that the war would soon have consequences for their own well-being and security, particularly through the threat of terrorist attacks.

Adapted from Hörschelmann (2008b)

12 and 18 years of age, they conducted focus group discussions, short questionnaires and interviews as well as using diaries, photography and other methods in order to engage with young people's views and experiences.

Drawing upon their work with two groups of young women who

constructed mental maps of their everyday lives, Hörschelmann and Schafer (2005) noted that for one group of young women, the local youth club was a central focus with the other places they use also being very close to their homes. As such, these young women only tended to engage with the global through shopping or through media consumption. These young women lived in an area dominated by 1980s high-rise buildings. Their photographs also show the local connections in their lives, with most showing their friends, parents or siblings. Furthermore, the images of their bedrooms did not appear to show many personal items or posters and those photographs that did show the shopping centre or such like only displayed indicators of globalisation implicitly rather than explicitly. However, Hörschelmann and Schafer (2005) did not simply conclude that global forces do not influence the young women's local lives. Instead, they explained that one of the main reasons why the girls tended to hang out close to their homes was due to the threatening presence of foreigners in other parts of the city which led the girls to withdraw from or avoid such places. 'As a result, they experience global culture primarily through education, media consumption and local shopping, rather than as a part of a daily use of urban space' (Hörschelmann and Schafer 2005: 231).

This contrasts with the second group whose mental maps included a range of places across the city with their homes occupying the margins. Although focused strongly around family members and friends, travel routes were a key feature of these maps. This group lived in the south of the city which had been gentrified. All maps show connections external to the city whilst also indicating regular contact with places where the signs of globalisation are evident: the cinema, language classes, the Bagel-café or shopping in the city centre. Some young women's maps also include what Hörschelmann and Schafer (2005: 233) refer to 'dreamscapes' which are their hopes for the future, including a desire to travel to particular countries and places. The girls' photographs also reflect these interests, with parents and family taking a less central role compared with images of them engaging in activities with others. These girls live in an older part of town, indicating their higher social status compared with the girls in the first group. The photographs of the girls' bedrooms included a mixture of postcards and posters reflecting their interests and their favourite pop stars. Two of the girls' photographs also show their computer and television as well as books. One young woman – Karin – also photographed herself and friends attending protests against the Iraq war. Unlike the first group, these girls often visited other parts of the city for shopping or leisure.

As this example demonstrates, globalisation has differential influences on the local lives of young people. Some young women reject global forces whilst others are open to what globalisation offers them. The first group showed some resistance and tended to focus on their immediate neighbourhood, whereas the second were eager to engage with global issues and imagine future access to a range of global resources. Overall, Hörschelmann and Schafer (2005: 239) conclude:

> On the one hand, we hope to have shown that young people, like the girls in this study, are differentially positioned in the local/global networks that, to a large extent, set out their present and future life options. On the other hand, however, we have sought to demonstrate the that young people, through their everyday socio-cultural and spatial practices, can incorporate the global very literally into their personal identities and/or reject it by drawing a tight line around their daily activity space.

Hörschelmann and Schafer (2007) also explored how young East Germans encounter and engage with mediascapes and ethnoscapes identified by Appadurai (1996). According to Appadurai (1996: 35) mediscapes relate to the global circulation of images:

> These images involve many complicated inflections, depending on their mode (documentary or entertainment), their hardware (electronic or preelectronic), their audiences (local, national or transnational), and the interests of those who own and control them. What is most important about these mediascapes is that they provide (especially in their television, film and cassette forms) large and complex repertoires of images, narratives, and ethnoscapes to viewers throughout the world, in which the world of commodities and the world of news and politics are profoundly mixed.

The young people involved in Hörschelmann and Schafer's (2007) research were most likely to engage with global mediascapes through television and radio as internet access tended to be restricted depending on parental income or on parents granting permission. For the young people involved, television remains the main mechanism of access to different lifestyles and images, and many identified Anglo-American television programmes as the most popular. The young people often talked about the idealistic images of America presented through television programmes with some criticising such representations in order to distinguish themselves from others who viewed such programmes in a less critical way. Intergenerational relations are important here as some parents encouraged their children to watch news programmes and documentaries emphasising their importance as an educational tool. Hörschelmann and Schafer (2007: 1864) argue that such programmes extend young people's 'geographical horizons and lead to greater knowledge of distant peoples and places' leading to some young people being very aware of geopolitical issues and contexts. At the same time, however, less well-educated young people were less confident about knowing where particular places were and so may have more restricted trajectories.

Experiences of travel were one of the biggest differences between the young people with those from poorer backgrounds having much less experience than their richer counterparts. Socio-economic differences between the young people were one of the main factors in determining particular resources and experiences. Those young people who did travel abroad tended to experience other places as tourists during family holidays with some of the older young people enjoying cheap trips to Eastern Europe with friends. They regarded their experiences of travel as being a very normal aspect of their lives. Many of the young people associated the countries they visited with poverty and deprivation compared to what they explained as their privileged backgrounds. Hörschelmann and Schafer (2007) argue that the experiences of the young people who are able to access travel opportunities will benefit them in the long term due to the language skills and social networks they engage with in the process. At the same time, some of the young people displayed xenophobic and racist attitudes towards different cultures or groups of people, with Turkish migrants being the subjects of hatred and fear. Overall, young people from all backgrounds were very interested in travelling or in relocating. They had desires, or what Hörschelmann and Schafer (2007) call 'dreamscapes' but varied according to the confidence they had in order to achieve this. The researchers concluded 'young people's immersion in the globalising world is highly uneven and depends strongly on where there are places in relation not just to economic but also to sociocultural resources' (Hörschelmann and Schafer 2007: 1869). Furthermore, they also point to age and gender as being important factors in shaping how young people negotiate and encounter the global.

Focusing upon the everyday lives of young women in the UK, Kehily and Nayak (2008) have also explored the ways in which young people respond to globalisation (see Box 8.3). In particular, they focus upon consumption practices associated with the media, clothes, the Internet and the ways in which young people's consumption practices shape and are shaped by global processes. They observe that for young women, globalisation may offer opportunities for new forms of femininity to emerge as they take advantage of changing patterns of work and consumption and can overcome the limitations of 'the patriarchal past' (Kehily and Nayak 2008: 325). Like the work of Hörschelmann and Schafer (2005, 2007), an important theme in this work is the inequalities negotiated by the young women based on the ability to consume and to be mobile. Middle-class young women may have access to the global market place whilst working-class young women are confined to their local communities, and are restricted in their access to particular educational and employment opportunities. However, Kehily and Nayak (2008: 330) show some optimism, noting:

> [I]t is possible to suggest that some cultural characteristics may be highly valued in the symbolic economy to be successfully utilised by working-class subjects. It is also possible that high exchange value may be attached to 'hot' qualitities and attributes that traverse class boundaries such as beauty, style, sporting ability, musicality or ICT wizardry.

Overall, Kehily and Nayak (2008) found that global media cultures were important aspects of young women's lives and were crucial for them as they sought to create relationships with others, construct and contest their identities and seek to understand their place in the world (Box 8.3).

Global inequalities

A clear theme that emerges from work about the perspectives on the relations between young people, place and identity is the multiple sets of relationships that exist between the global of other scales, in particular, the local. Globalisation influences young people's daily lives through a variety of means and at a range of spatial scales, such as through social institutions, urban spaces and community encounters. Young people negotiate different global images, signs, texts and media

Box 8.3

Global femininities: film, music and electronic media

Kehily and Nayak (2008), drawing upon comparative ethnographies in English state schools, explored young women's responses to globalisation. They considered globalisation from below by investigating young women's connections with the global and the ways in which global media culture were important aspects of the young women's lives. Pointing to the importance of schools as features of young people's daily lives – where they come into contact with the 'products of cultural globalisation' (Kehily and Nayak 2008: 326), they demonstrated the importance of 'local–global negotiations to the making of gender' (p. 326):

> As high consumers young women regularly interact with a global bricolage of media signs, commodities, music, film and magazines. These global products have a bearing on who they are and how they wish to present in school and neighbourhood cultures.
>
> (Kehily and Nayak 2008: 327)

In their work, they emphasised the importance of recognising that 'young women . . . are not only positioned by cultural commodities, they are engaged in the representational sphere and produce meanings for themselves' (Kehily and Nayak 2008: 329). In doing so, they focus upon three aspects of global cultural media: film and soap operas; music and dance; and electronic media and new technologies.

Film and soap operas

For Kehily and Nayak (2008), film and soap operas are products of cultural globalisation that offer young women access to different perspectives on the performance of new femininities. The young women they worked with discussed popular prime-time soaps such as 'Home and Away' and 'Neighbours' – which were identified as having a safer content aimed at children – and 'Coronation Street', 'Emmerdale' and 'Eastenders' – which tended to include more sensitive adult-focused themes such as drugs, sex and sexuality. For Kehily and Nayak (2008: 332), these soap operas 'offer spaces in which new femininities can be performed through the relatively safe mediated genres of television fiction', and so their engagement with global media is rooted in their local everyday lives. Furthermore, through engaging with these different soaps and television programmes – enhanced through cable and satellite – 'collective viewing or discussion offers an opportunity for femininities to be produced, defined and enhanced' (Kehily and Nayak 2008: 333).

Music and dance

Although work about globalisation argues that global flows of music, style and media offer young women the opportunity to engage with forms of identification and belonging beyond the local, Kehily and Nayak (2008) argue that their ethnographic work shows that place still matters in terms of how young women construct and contest their identities. They found that global forms of music were consumed in meaningful ways by young women and actively used in the negotiation of their gendered identities. In particular, young women regularly discussed the looks, talent and general behaviour of global pop stars which were materialised through music, magazines, clothes, videos, DVDs and other such forms of global media.

Electronic media and new technologies

'The prevalence of website and personal blogs suggests that young women have new media resources with which to develop identities and social networks' (p. 337), meaning that young women's personal space, such as their bedrooms, now enables them to access various global resources and networks. They can access virtual spaces such as MySpace, Facebook or Bebo in order to talk to friends, share experiences and articulate their feminine identities.

in the process of creating and sustaining their own youth cultures. For Wyness (2006: 69):

> Globalisation has undoubtedly become a powerful economic, political and cultural frame of reference. The rise of the individual and the emphasis on choice encourage children and young people to have an active role within adult settings as cultural and economic agents. However, globalisation also offers a broadening of perspective on the way we see the world. At the same time as becoming global consumers, we are also likely to make connections between our own local and national situations and the situations of others in quite different national cultural contexts.

Furthermore, although there are complex global processes and transformations taking place, 'the social reproduction of children and youth, by contrast, takes place at the local or national level at best and historically has often been relegated to the private and/or domestic sphere' (Ruddick 2003: 334). A strong focus in debates about globalisation is the ways in which global forces are lived out, experienced and felt locally. Indeed, many would argue that in order to understand global issues, we need to appreciate how these operate locally:

> If we are to understand the local nature of our lives, including the local expression of youth lifestyles, we have also to consider the global context within which the local operates . . . We can only

> understand the changes taking place in our locality if we are prepared to understand changes taking place outside our locality that may directly or indirectly affect the nature of that locality
>
> (Miles 2000: 60–1)

Drawing upon his work with young people in the north-east of England, Anoop Nayak (2003a: 170) concludes that although many of the trends associated with globalisation – international migration, technological progress and cultural exchange – are not necessarily new, 'there has been an acceleration and intensification in these processes to the extent that most lifestyles are now inextricably tied into a web of global relations.' In focusing upon the ways in which the global is lived out locally, Nayak (2003a: 5–6) challenges the overall authority of the global when he observes:

> Empirically grounded place-based analyses of young lives may now offer a challenge to wider perceptions of globalisation as an omnipresent, homogenising force that goes unheeded, in favour of a more textured and contingent portrayal of youth cultures.

As well as focusing upon the locally experienced outcomes of globalisation, important work about young people, place and identity has also drawn attention to the ways in which globalisation is experienced in similar ways in disparate places. Drawing upon the experiences of young people in New York and rural Sudan, Cindi Katz (2004) is one of the few scholars who has focused upon the ways in which global economic restructuring has influenced social reproduction. A systematic lack of investment in New York and other urban areas of the USA has contributed to the decline of neighbourhood and community. Alongside this, issues such as gang communities, teenage pregnancy and unemployment have all worked against the process of improving the situations of young people. In Sudan, Katz found that young people were also experiencing similar processes of exclusion and marginalisation. Young people were working long hours in a variety of subsistence activities, yet were unlikely to have access to productive land when they come of age because the demographics of Sudan mean that there are too many families competing for small quotas of land. This could result in many young people being excluded from earning a wage when they become adults, despite their experience and knowledge of farming.

Just as globalisation influences the lives of young people possessing a diverse range of social identities, there are also significant concerns about how it shapes the lives of indigenous young people. As part of the Oxfam International Youth Partnerships Youth Commission Report on Globalisation, and drawing upon submissions made to the commission by indigenous young people, Bruce (2003) identifies some of the main issues that globalisation raises for the rights of such young people. First, globalisation is seen to increase the likelihood of indigenous young people being displaced and so raises concerns about access to land. Second, the structure of the global financial system is seen to negatively impact on indigenous youth as the power and control of multinational companies sideline, overlook and ignore the perspectives of indigenous communities. Third, although developments in communication technologies are often seen as a positive spin-off of globalisation, a lack of consultation with indigenous young people may lead to them being marginalised as technological developments work against specific aspects of their cultural heritage. A fourth concern relates to the commercialisation and privatisation of education that may restrict the access of indigenous young people to appropriate education and exclude indigenous languages. Finally, Bruce (2003) points out that indigenous young people lack appropriate representation in decision-making processes that affect them and are often excluded from international decision-making bodies and forums.

Key themes

- Research about globalisation has increasingly challenged the notion that the global is all-encompassing and has instead drawn attention to the ways in which global issues are lived out locally and reworked and recreated in different places in different ways.
- Global events often influence how young people construct and contest their identities as processes of exclusion, marginalisation and discrimination are intensified in light of the ways in which particular groups are demonised following such events.
- Young people's local engagements with global issues are complex and multifaceted, connected as they are through a range of media and popular culture such as television, pop music, magazines and the internet.
- Globalisation has worked to increase inequalities between young people from different parts of the world as well as also marginalising young people in similar ways in very different places.

Project ideas

Construct a mind map of your weekly activities, thinking in particular about the things you do, where you go, who with, how you get there and why. After doing so, think critically about the ways in which these activities are informed or shaped by global issues.

Discuss the ways in which global geopolitical issues shape and are shaped by young people's everyday lives and social identities.

To what extent do you consider young people's processes of identity construction to be global in character?

Suggested further reading

This article charts the relationships between childhood and other social transformations such as globalisation:

Aitken, Stuart C. (2001) Global crises of childhood: rights, justice and the unchildlike child. *Area* 33(2) 119–27.

This text explores the cultural dimensions of globalisation and includes discussion of key global cultural flows:

Appadurai, Arjun (1996) *Modernity at large: cultural dimensions of globalisation.* Minneapolis: University of Minnesota Press.

This excellent article provides a rich account of young people's responses to the war in Iraq:

Hörschelmann, Kathrin (2008) Populating the landscapes of critical geopolitics – young people's responses to the war in Iraq (2003) *Political Geography* 27(5) 587–609.

This article explores how the lives of young people growing up in eastern Germany are shaped by globalisation and how the young people respond to these processes:

Hörschelmann, Kathrin and Schafer, Nadine (2007) 'Berlin is not a foreign country, stupid!' – growing up 'global' in Eastern Germany. *Environment and Planning A* 39 1855–72.

This is an excellent book that provides a rich and textured ethnographic account of the place of economic restructuring in young people's lives:

Katz, Cindi (2004) *Growing up global: economic restructuring and children's everyday lives.* Minneapolis: University of Minnesota Press.

This excellent article explores the ways in which global media and consumption are interconnected with young women's place-based practices:

Kehily, Mary Jane and Nayak, Anoop (2008) Global femininities: consumption, culture and the significance of place. *Discourse: Studies in the Cultural Politics of Education* 29(3) 325–42.

Part III

Themes and sites

9 Institutions

Gavin: In the first year I felt the same as the other lads, there were six of us, real good mates and we just hung out and did stuff really, but by the second year they started getting into lasses and I didn't, and that's when the bullying started, getting called 'puffter,' 'queer' and stuff like that, real nasty stuff. (Casey 2002: 67)

Trisha: I've tended to keep friends from back home, you know, you've got your few select university friends that you would call friends rather than . . . acquaintances. You know [people] you get on with, do your group work [with], you'd go to the pub and things like that, but regards going out at weekends and that sort of stuff it tended to be people back home . . . friends that you've been friends with for years, and you know, they're going to be there once you've finished university . . . when everyone else has gone away back home. (Christie 2007: 2456)

Sharon: It's a fine place to be, better than being in foster care because there is more kids here your own age that have been through the same sort of thing. (Edmond 2003: 328)

Young people's institutions

The views of Gavin, Trisha and Sharon all draw attention to the multiple ways in which young people negotiate different institutional contexts on a daily basis. These negotiations of place, power and boundaries are complicated further by the ways in which different identities are actively communicated, interpreted, variously negotiated, represented and performed in ways which shape and are shaped by different institutional settings. Gavin reflects upon managing his sexual identity at school, Trisha discusses how she feels about her peer group given her transition to studying at university and Sharon talks about her views about living in residential care. There are, however, many other institutional settings relevant to young people, place and identity: after-school care, college, religious institutions, prisons, special schools and summer camps (Thomas 2000). There is not enough space to consider all of these institutional settings in this chapter, so I focus attention on school, university and residential care to provide three sets of examples of the intersections between young people, identities and institutions. There is much youth research, particularly within the youth transitions tradition, which focuses on young people's negotiations of the transitions through school to university, work and increasing independence (Heath *et al.* 2009, MacDonald and Marsh 2004, MacDonald *et al.* 2001) and this scholarship is important in understanding young people's negotiations of different institutional contexts.

It may appear somewhat strange having a chapter about institutions in a book about young people given that 'youth' is sometimes thought of as an institution in itself. Ruddick (2003: 336) points out that 'modern "childhood" and "youth", as strictly age-graded phases in the life cycle, require supervision and training in an array of institutional contexts, from schools to camps to recreational programmes', and so it is crucial to consider different institutional contexts in order to gain a full appreciation of the intersection of young people, place and identities. Furthermore, 'institutions, which to a certain extent both stand in for, but also stand apart from the home, have received relatively little attention from geographers, despite the fact that they constitute an intriguing type of social arena' (Valentine 2001: 141).

Philo and Parr (2000: 513) have explored the 'intersection of the institutional and the geographical' differentiating between the geography of institutions which refers to the relative spread of particular types of institutions and geography in institutions, highlighting the importance of considering the internal spaces, negotiations and contestations within different institutional contexts. As Philo and Parr (2000: 514) observe, it is important to think critically about how institutions are 'practically and conceptually shaped in many different ways'.

> Institutions have usually referred to those material built environments . . . which seek to restrain, control, treat, 'design' and 'produce' particular and supposedly improved versions of human minds and bodies. Big buildings with large grounds, lots of rooms and corridors, and sizeable resident populations: these are the focus of those institutional geographies which have previously caught our attention.
>
> (Philo and Parr 2000: 514)

Evans (2007: 189–90) draws upon the idea of the 'total institution' that was developed by Goffman (1961) and identifies four key characteristics of institutions. First, there is 'batch living: this refers to block treatment of residents, with no opportunity for personal choice regarding clothes, food or personal space'. Second, 'binary management: this refers to the social distance between staff and resident; they eat separately (possibly with different menus) and pursue leisure activities separately, with the staff supervising rather than participating. Power is exerted through this separation of worlds.' Third, 'the inmate role: the resident is stripped of their former identity and becomes, essentially, depersonalised. Everyday life is organisation centred rather than user centred'. Finally,

> the institutional perspective: over a period of time the institution completely takes over and becomes the only frame of reference for the individual, who cannot see beyond the institution and more or less succumbs to its demands. The rigidity of routine and lack of power are accepted. This feature describes the process of becoming institutionalised and may apply as aptly to workers in large organisations as it does to patients in a psychiatric hospital.

Negotiating school

Schools are crucially important contexts for young people. For many, they have profound influences over how young people feel about themselves and their multiple identities, and who young people become as adults. As Casey (2002: 63) notes, 'though the process of school based identity production can never be final, it can have long and significant consequences on individual lives'. The places within and related to school are important aspects of everyday life for many young people. They are the crucial sites for how young people construct their identities, the mechanisms whereby identities come under surveillance, and for the ways in which identities are given meaning. 'Of the spaces outside of the home, schools in particular have long been researched for their potential to encourage – or to damage – a student's self-concept, actual achievement and motivation to work toward future goals' (Krenichyn 1999: 44). Places within schools do not exist in a pre-given form and are instead relational and in the process of being made. According to Valentine (2001: 142), 'two worlds make up the school. First, there is the world of the institution. This is the adult-controlled formal school world of official structures: of timetables, and lessons organized on a principle of spatial segregation by age.' Second, there are the young people's informal worlds, friendship groups and social experiences. Morrow (2001: 44) observes that 'from the perspectives of young people, school is an important "community" in its own right, and each school has its own specific culture and environment', and Hyams (2000: 635) has noted how schools are 'constitutive of and constituted by social relations or power.'

McGregor (2004) observes that schools are public institutions that are often structured and designed in particular ways. Many schools have classrooms, assembly halls, playgrounds and laboratories, and we rarely question their structure or order. Surprisingly little research has been conducted about the built environment and places of schools or the interaction between pupils in these spaces. Yet, schools are clearly very significant locations in the everyday lives of many young people and are places that are imbued with power relations and controls. Through the control of time, schools exercise influence over the distribution of resources, bodies and space. Furthermore, the implementation of school rules and regulations often involves the monitoring and control of particular spaces, such as entrances and exits, and rooms being out of bounds, such as the staffroom. Schools are therefore a place where young people learn much about the dominance of adults in society and the way in which power and control operate in society.

Casey (2002) has studied the experiences of young gay men at school and how this relates to broader issues about their sexual identities and the process of coming out:

> For the school pupil, encountering the sexuality of students and teachers is part of the school experience and initial sexual encounters often occur during this time. The school culture expects such interactions, and will often promote them through school discos, school paper gossip columns, sports groups, sex education, the yearbook and so on.
>
> (Casey 2002: 63)

Schools are important in reinforcing norms and practices around acceptable forms of identity, particularly those associated with the performance and articulation of gendered and sexualised identities. As Casey (2002) observes, it is in school that many young people learn that heterosexuality is the norm and that everyday spaces are encountered as heterosexual. The period when young people explore their gendered and sexual identities can be very stressful and challenging for gay young people when they may experience bullying or other forms of marginalisation. Casey (2002) found that some young gay men started making observations about other young men's bodies when they were ages 12 or 13, without necessarily seeing this as something directly related to their sexual identity. For others, it was not until they were in their late teens or early twenties that they became aware of their sexual feelings. For those who are aware of their sexual feelings whilst at school, the dominant culture of heterosexuality can work to increase the likelihood of them experiencing homophobic bullying, along with having their behaviour policed regularly.

By the time many young men leave the heavily heterosexual environment of the school, they are often ready to come out to themselves and to their friends:

> *Lee:* Well I was about 17 at the time, I told my best mate. She said she knew anyway and was cool with the whole idea, she even came on the scene with me.
>
> (Casey 2002: 71)

Mac an Ghaill (1997: 51) has conducted influential work about the ways in which 'heterosexual male students develop a mode of masculinity in relation to the social structure of the secondary school.' In doing so, he explored the ways in which different forms of masculinity and different ways of being a male student are articulated in the school environment. He found that students' relations with their families, their experiences of the school and local labour markets were all crucial in the ways in which masculine identities were performed and articulated. Through his observations of the practices, behaviours and attitudes of young men at school, he identified a number of working-class heterosexual peer groups in the school environment: the Macho Lads, the Academic Achievers, the New Enterprisers and the Real Englishmen (see Box 9.1). Classifying young men into this typology, Mac an Ghaill (1997) acknowledges that the boundaries between these groups are flexible and are therefore not fixed categories; however, they represent one way of understanding the masculine

Box 9.1

Masculine peer groups in school

The Macho Lads were all in the bottom two sets for all of their subjects. They regarded school with hostility and experienced much of it as meaningless. Teachers saw them as the most anti-school group of young people. Things that are important to them were, 'looking after your mates', 'acting tough', 'having a laugh', 'looking smart' and 'having a good time'. Their masculine identities tended to focus on issues of territorial control, solidarity and physicality. As Mac an Ghaill (p. 58) notes, 'the Macho Lads rejected the official three Rs (reading, writing and arithmetic), and the unofficial three Rs (rules, routines and regulations)' and instead opted for the three Fs – 'fighting, fucking and football', seeing school as an 'apprenticeship in learning to be "tough"'. They strongly disliked what they saw as the authoritative nature of teachers and associated schoolwork with femininity and 'dickhead achievers' (p. 59):

> *Leon:* The work you do here is girls' work. It's not real work. It's just for kids. They [the teachers] try to make you write down things about how you feel. It's none of their fucking business.
>
> (p. 59)

The Academic Achievers were a small group of young men who had a generally positive relationship with the curriculum, and comprised a number of Asian and white young men from skilled working-class backgrounds. Some of them became associated with arts subjects and were often positioned by teachers and pupils as 'effeminate'. As they became increasingly aware of their gendered positioning within the school environment, they often responded creatively by subverting dominant institutional practices associated with gender and sexuality. These young men were eager to pursue a professional career, however, their identities relied upon a strong working-class work ethic, unlike the middle-class success of the Real Englishmen.

The New Enterprisers were inclined to participate in mini-enterprise groups, driven by their enthusiasm to obtain high-skilled jobs when leaving school. They responded to the emphasis upon 'values of rationality, instrumentalism, forward planning and careerism' (p. 63), drawing upon the focus upon the obtaining technological and vocational skills whilst at school:

> *Wayne:* In class you just sit there in most lessons as the teacher just goes on and on, but in business studies and technology, you learn a lot, you're really doing something, something that may be useful for your future.
>
> (p. 64)

> The Real Englishmen displayed an ambivalent relationship to the school curriculum. They refused to accept the authority of teachers, though their resistance was more individualist and varied than that of the Macho Lads. These young men had plans for higher education and a professional career and positioned themselves as the 'arbiters of culture' (p. 65), evaluating teachers and students in terms of their cultural capital, building on their publicly confident masculinity which tended to over-emphasise the significance of cultural capital. These young men emphasised 'honesty, being different, individuality and autonomy' (p. 66), and strongly disliked the Macho Lads.
>
> Adapted from Mac an Ghaill (1977: 58–66)

identities of young men being studied. This work emphasises the relationality of identities and group practices and Mac an Ghaill (1997: 61) highlights the importance of considering not only 'gender differences but also relations between young men and women and within young men's peer groups.'

Hyams's (2000) work with young Latina girls in Los Angeles explains how the girls, having made the transition to high school, experience the process of becoming sexual subjects that are both 'desirous of and the object of desire in intimate relationships with boys' (2000: 640), yet such practices are controlled and monitored by those in authority, particularly parents. A key concern for these young women is how they manage their bodily comportment when in the company of young men at school, and therefore how they present themselves as young women rather than as girls. There is a sense, then, that the young women have a choice between being a girlfriend and being an academic achiever. Overall, Hyams (2002) highlights that gender relations and dominant discourses associated with gendered and sexual norms both in and out of school have strong influences over the young women's everyday behaviours. The gendered dimensions of spaces within school also emerge in Krenichyn's (1999) research in a high school in New York. In particular, the school gym is a regular focus of conversation amongst the young people where young women are marginalised by the boys' dominance of the gym.

Managing university

Universities are also institutional contexts that tend to be dominated by young people and recent years have seen a rapid expansion in university provision across the world. Many cities now include large groups of students to the extent that some British cities have student populations of around 10 per cent of the total population (Chatterton 1999). There are large numbers of traditional students studying at particular universities, yet the increase in student numbers is largely due to the fact that many students from non-traditional backgrounds are now attending university (Smith 2009). As such, attending university is no longer an experience that is open only to young people from particular socio-economic backgrounds who possess particular identities. The mix of students at any university can, however, play a powerful role in shaping students' identities and senses of belonging within the university environment. Furthermore, the reputation of a university, coupled with the stereotypes associated with particular higher education institutions also influences and is influenced by the young people who study there. Alongside this, the institutional context of the university has an important role in shaping student identities. The importance of student halls of residence, the lecture theatre, the library and other university activities all work to enforce the identity of being a student upon young people, separating them from those who are not a part of this institutional framework.

The transition to university is a key phase in the lives of many young people. 'Not only does this transition often require the use of a different institutional space, but it also involves a series of negotiations and decisions' (Hopkins 2006: 244). Some of the main concerns that influence students' decisions about where and what to say include examination performance, family expectations, and previous educational experiences as well as money, loans and debt and accommodation, housing and moving away from home. Working with a group of students, many of whom were the first in their families to consider attending university, I have explored the hopes and fears of students who were about to experience the transition to university. In terms of the hopes about university:

> Consistent across the three groups was the importance of a sense of achievement, the potential for new social circles and peer groups, and the chance of getting a good job at the end of their studies. These success-oriented comments were also balanced by much discussion about the potential for a new social life accompanied with drinking, clubbing and 'nights out'.
>
> (Hopkins 2006: 244)

Focusing upon the experiences of traditional students studying at Bristol University in the UK, Chatterton (1999) draws attention to the relationships between university students and different spaces within the city. He gestures towards the important role often played by family and intergenerational relations in the formation of student identities, as parents may encourage their offspring to study at particular universities and/or to focus their attention on particular types of qualifications. Chatterton (1999) also discusses the ways in which university campuses work to create clusters of students around universities at particular times of the year. As well as congregating around universities, traditional students also have a wider impact on the city as they occupy pubs, clubs, cinemas, music venues, exhibitions and such like. Students often spend much time in neighbourhoods and communities near to the university and in which they live. Their use of these spaces is often learnt from other students, institutionalised through freshers' week and promoted through advertising and other independent publications. However, Chatterton (1999) also argues that as students progress through university, many seek to distance themselves from mainstream life by trying out new and different places, pointing to the ways in which the relations between students, place and identity change as students pass through the university system. This is supported by Smith and Holt (2007) who observe that as students' progress through university, they tend to move into the private rented sector and share with friends of their choice. Students have some autonomy in choosing where to live and who with, often wanting to live in areas with particular social and cultural resources and facilities.

Smith and Holt (2007) draw attention to the relationships between higher education students and urban change in the UK, arguing for the need to consider gentrification from a lifecourse perspective. They observe how much work about gentrification has not considered the experiences of young people and students despite the fact that young people have often played a key role in redefining, reimagining and redesigning urban spaces. 'Since the mid-1990s increasing numbers of students have moved into distinct enclaves of university towns and cities' (Smith and Holt 2007: 147), with 150 wards in the British census containing a student population in excess of 20 per cent of the total population. This pattern of residential segregation has been heightened by the creation of student areas by different public and private sector organisations (e.g. universities, property developers and local government). According to Smith and Holt (2007: 148), 'it can be argued that the commodification of student spaces or lifestyles is underpinned by specific unfolding structural conditions. Important factors here include: the state-sponsored expansion of higher education to foster global economic competitiveness . . . decreasing welfare provision for higher education students' along with increasing demands for higher education due to higher expectations of society and the changing nature of housing provision in the private sector (see Box 9.2 and Table 9.1).

Smith and Holt (2007) found that over 80 per cent of first-year university students at the universities of Leeds and Brighton selected university-maintained housing as their first preference for accommodation. They see this as part of students' strategies for coping with the move to university:

> Empirical findings from Leeds and Brighton suggest that many students often have an ambiguous sense of self as adult or nonadult

Box 9.2

The four dimensions of studentification

Recognised as being developed by Darren Smith, 'studentification engenders the distinct social, cultural, economic and physical transformations within university towns, which are associated with the seasonal in-migration of HE students (Smith 2005: 73). It is regarded as having four main dimensions:

Economic:	the revalorisation and inflation of property prices, which is tied to the recommodification of single-family housing or a repackaging of private rented housing to supply housing in multiple occupation for HE students. This restructuring of the housing stock gives rise to a tenure profile which is dominated by private rented, and decreasing levels of owner-occupation.
Social:	the replacement or displacement of a group of established permanent residents with a transient, generally young and single, middle-class social grouping; entailing new patterns of social concentration and segregation.
Cultural:	the gathering together of young persons with a putatively shared culture and lifestyle, and consumption practices linked to certain types of retail and service infrastructure.
Physical:	associated with an initial upgrading of the external environment as properties are converted to housing in multiple occupation. This can subsequently lead to a downgrading of the physical environment, depending on the local context.

Adapted from Smith (2005: 74–5)

Table 9.1 ***The effects of studentification***
Impacts on the social, economic, cultural and physical dimensions

Social	*Economic*	*Cultural*	*Physical*
Demographic structure of the local population Levels of population density Levels of population stability/transience Turnover of residents/ property Cohesion of local community and community interaction Levels of neighbourliness Meaning and symbolism of location Supply and demand for schools and local health services Supply and demand for public transport Effectiveness of crime prevention strategies and self-policing Trends of criminal activity Levels of electoral voting and political affiliations Effectiveness of car parking schemes and provision Strength of local voluntary schemes/sector Levels of alcohol/drug abuse Health and well-being of local people	Supply and demand for housing Buoyancy of housing market Portfolio of housing stock Flexibility of housing stock Supply and demand for affordable housing Condition of housing stock Spending levels within local economy Levels of inward capital investment Supply and demand for services of letting/estate agents, property maintenance and building contractors Supply and demand for local retail, leisure and recreational services Seasonality of local economy and services Levels of housing abandonment Supply and demand for domestic services Supply and demand for childcare services Levels of council tax revenue Local workforce	Supply and demand for specific leisure, recreational and retail facilities Levels of antisocial behaviour Levels of noise nuisance from households, pedestrians, taxis/ private vehicles (In)compatibility of lifestyles Supply and demand for levels of policing and emergency services	Levels of private vehicle use and cycling/walking Levels of traffic congestion Levels of visual pollution (to-let signs) Effectiveness of refuse and waste collection Levels of litter and rubbish Upkeep of gardens and driveways Upkeep of external environment Levels of graffiti and vandalism

Adapted from Smith (2005)

> during this phase of the lifecourse. Studenthood can therefore be viewed as a liminal period for young people, in which the boundaries between independent adulthood and dependant childhood mesh, and often conflict. Within institutional spaces perceived as 'safe' and 'supportive', students are able to adjust and reconcile 'everyday' stresses associated with their newfound independence, such as time management, control of finances, and personal space.
>
> (Smith and Holt 2007: 151)

Furthermore, by living alongside other students, they also have direct access to a social life and other aspects of student lifestyle. Smith and Holt (2007) also draw a number of connections between the geographies of students and those of young professionals and recent graduates, particularly with reference to residential patterns and housing choices. They suggest that the boundary between student lifestyles and those who have recently completed their studies are likely to become increasingly blurred and so may result in the need to question the uniqueness of studenthood as a distinct phase of the lifecourse.

Alongside the socio-economic issues relating to the processes of studentification, it is also important to consider the internal places, structures and processes that operate in university contexts to reinforce particular identities whilst stigmatising or marginalising others. As well as the structures of lectures, tutorials and seminars, there are also more subtle processes associated with membership of particular student groups, social background and degree subject which may work to bolster particular identities and marginalise others (see Box 9.3).

Box 9.3

The 'Coffee House' in McGill University Law Faculty

'At 4pm nearly every Thursday afternoon during the teaching year at Montreal's McGill University, between 100 and 300 law students (well over one third of the total student body) start to crowd into a recently built space designed of steel and glass called the "Atrium" in the Law Faculty. Half the time this event is sponsored by Montreal and Toronto law firms who send their youngest and best dressed lawyers to mingle with the crowd and convince those to whom they speak that their firm is the most successful, the most prestigious and the highest paid' (Turner and Manderson 2007: 761). Turner and Manderson (2007) were interested in exploring how students experienced this particular social space within the university context and did so through the use of individual interviews, informal discussions and participant observation. Turner and Manderson (p. 768) argue that the power of this space lies in the ways in which it 'presents a certain model as central and marginalises alternatives.' It promotes a certain image of 'real law' and so normalises particular practices, attitudes and values, plus it operates to create an in-crowd who attend the events and a crowd who are on the margins of the community. The boundaries of the group were also strongly monitored and policed with gate-crashers being classified as 'SNAILS' (Students Not Actually In Law School) (p. 769).

According to many of the students consulted, their attendance at Coffee House events was purely for social purposes and the participant observation of the researchers found that most students focused their attention on obtaining drinks from the bar and talking to their friends. However, Turner and Manderson (p. 773) also observed a small group of students who actively participated in networking with the lawyers present at the event. These students were often in their final year of law school and often dressed-up for the occasion and were identified by the researchers as consciously attempting to 'work the room'. Only a minority of students admitted to such practices although many were aware of others engaging in such tactics. Many publicly ridiculed networking and even established particular behaviours and spatial practices in order to avoid having to speak to any of the visiting lawyers.

Although a majority of students attended the event only to drink fine wine and eat good food with their friends – and actively resisted networking with lawyers – Turner and Manderson (2007) argue that the practices of the students were all about behaving and acting like lawyers, even if only with a small group of friends. They argue that 'these repeated performances represent an embodied notion of what it is to be a McGill law student on his or her way to a career as a successful corporate lawyer. This identity is constantly being reinforced through repetition, week after week' (p. 775). They continue, 'Coffee House, as a site of repetition, operates to create a recognisable identity – a powerful corporate lawyer – whose legitimacy and prominence call forth its many aspirants' (p. 775) and 'student identities are slowly transformed, metamorphosed, little by little, as Coffee House socialises those who attend' (p. 779).

Turner and Manderson (2007)

Residential care

As this chapter is demonstrating, young people engage with a variety of different institutions in their everyday lives and during their transitions to adulthood. Some children and young people may have to negotiate the institutional settings provided by residential care, often as a result of the absence or lack of appropriate care within the familial setting. There are approximately 10,000 young people in residential care in the UK at any one time (Kendrick *et al.* 2008). Many of these young people have experienced situations involving violence and various forms of abuse or are involved in the abuse of drugs or alcohol. They are often the most vulnerable in society and are likely to feel ashamed by being labelled with the identity of being in care. Young people in care often lack the intergenerational relations available to other young people during the transition to adulthood.

Gilligan (1999) identifies residential care as serving four key functions. First, there is the need to maintain young people's basic physical, psychological and emotional needs appropriate to their age and stage of development (maintenance). Second, there is a need to protect young people in residential care, particularly given the abuse and exploitation they may have endured before entering care (protection). Third, young people should be helped to recover from the previous deficiencies that led them into care. This might involve providing extra support in terms of education, health care or therapy (compensation). Fourth, young people should be prepared with appropriate practical skills and emotional well-being to leave care and continue to live a secure life (preparation).

Research by Gilligan (1999: 187) has drawn attention to the importance of considering the different components of young people's lives whilst in care:

> When one considers the domains in which a young person in care may live out daily relationships, they include family, care setting, school, peer group, neighbourhood, workplace, and leisure time interests and activities. Each of these domains is a source of potential relationships which may contribute positively to a young person's progress while in care.

Gilligan (1999) focuses on the leisure time of young people in care as a possible positive pathway in their journeys through and out of care. Participating in different leisure activities offers young people in care a chance to engage with mainstream youth culture. He suggests that leisure activities such as cultural pursuits (e.g. dance classes or singing in the school choir), the care of animals or participation in sport (e.g. basketball, running and football) can each help maximise the resilience of young people in care offering real benefits to them as they negotiate their everyday lives.

According the Edmond (2003: 322), 'the voices of children and young people in residential care have been somewhat muted', as there has been a tendency to focus upon the hierarchical relationships between young people and carers in residential settings, rather than exploring the relationships between the different young people who live in any one unit. Drawing upon ethnographic research in two local authority children's units in northern Scotland, Edmond (2003) found that many of the young people relied upon their co-residents during their everyday lives as well as important to helping them understand themselves and the social world. From this work, Edmond (2003) found that young people's sense of power and agency with the residential setting often related to their competence at dealing with particular social situations that arose within the unit. One such set of competencies or 'social currencies' (Edmond 2003: 327) is the provision of support and advice which was an important characteristic of everyday life for the young people in the units.

The sense of shared understanding and sense of support offered by the young people often related to the age and previous experiences of the young people with older young people sometimes offering advice or support to their younger peers. They often supported each other in stressful or upsetting situations or offered encouragement about school or work by 'sticking up for each other' (Edmond 2003: 331). There was also evidence of young people sharing their own possessions such as clothes, CDs and make-up. This was especially important for helping young people to develop a sense of identity given that many of them entered the residential setting with few possessions in the first place and it also helped them to build up a sense of trust through sharing resources. Other young people would celebrate birthdays by buying gifts for their co-residents. This really mattered for many young people given that their birthdays passed unnoticed and unacknowledged by family members due to the lack of intergenerational familial relationships:

> There was a sense of collective isolation from family and from society at large – a sense of being different from other groups of young people and so they would support each other and look out for each other. Boys often relied on support from other young people in dealing with external, often violent, threats.
>
> (Edmond 2003: 331)

Aside from the support offered by other young people, the adults working in residential care can be seen to take on a number of roles including being a caregiver, social worker, advocate, counsellor and mentor (Gilligan 1999).

Drawing upon research in two local authorities in England, Ince (2004) explored the experiences of young black people in care. She found that all young people were exposed to a 'Eurocentric model of care' (p. 218) and had restricted contact with their families and friends, with the black community in general or with their cultural background. 'The role of the family in transferring meanings, values, folkways, symbols and traditions was lost to those young people who were separated from their family of origin and community' p. 220). This resulted in young black people being distanced from their cultural heritage with a complete lack of opportunity to learn about cooking, how to care for their hair or skin or how to enjoy black art and history. According to Ince (p. 220), these young people were effectively 'stripped' of their identity and left the care system feeling embarrassed by their ethnic and racial identity. This is largely down to the failure of professionals and residential and foster carers to encourage young black people to feel any sense of identification with their cultural background. Furthermore, all young people reported experiencing racism whilst in care.

Ethical and methodological considerations

Conducting research in different institutional contexts requires an awareness of, and sensitivity to, the ways in which the practices, values, behaviours and attitudes of those within the institution influence how research takes place. The most obvious way in which such issues enter research encounters relates to the ways in which power relations work to empower particular individuals and disempower others. Read the following articles and think carefully about different institutional contexts where research with young people might take place. What are the key factors that should be considered when aiming to conduct ethical youth research in an institutional context? How might researchers manage such issues?

Holt, Louise (2004) The 'voices' of children: de-centring empowering research relations. *Children's Geographies* 2(1) 13–27.

Kendrick, Andrew, Steckley, Laura and Lerpiniere, Jennifer (2008) Ethical issues, research and vulnerability: gaining the views of children and young people in residential care. *Children's Geographies* 6(1) 79–93.

Key themes

- There are broad types of institutional contexts shaped, negotiated and contested by young people during their transition to adulthood and these range from institutional contexts that are informal and subtle through to settings which are hierarchical and overpowering.
- Arguably, school is one of the most fundamental institutions negotiated and contested by young people often playing a major role in shaping and being shaped by young people and the processes of identity construction. The values of practices of teachers and pupils create school environments that are gendered and sexualised in ways that are homophobic, sexist and dominated by specific forms of masculine behaviour. Specific locations within schools often become associated with particular social groups of young people to the exclusion of others.
- The transition to university is a key component of many young people's educational experiences. Specific universities are often stereotyped as being associated with particular types of student identities and particular types of subjects, influencing who does and does not attend and working to empower some students and marginalise others.
- Studentification refers to the processes whereby university towns experience social, cultural, economic and physical change during the annual influx of university students.
- Young people who lack appropriate familial inter- and intragenerational care are vulnerable and may have to enter residential care where peers are often – although not always – an important source of support, alongside the advocacy provided by staff working in residential care.
- The provision of residential care tends to focus upon particular groups of young people with particular identities, leading other young people to feel excluded and marginalised from developing particular connections with their cultural heritage or ethnic background.

Project ideas

- What are the main institutional contexts you negotiate each week and who are the different actors involved in these contexts? How do these contexts influence how young people construct their identities and in what ways have you shaped how such contexts are constructed?
- What are the different relations of power that young people negotiate within the school environment and in what ways might young people creatively shape such relations?
- What are some of the political and economic changes that have resulted in the growth of the university sector and what consequences does this have for young people's constructions of identities?
- Can you think of examples of institutional contexts that are designed, shaped and controlled by young people? What forms do these contexts take and how do they differ from other institutional settings?

Suggested further reading

This is a useful exploration of the relationships between university students and city centres:

Chatterton, Paul (1999) University students and city centres – the formation of exclusive geographies: the case of Bristol, UK. *Geoforum* 30 117–33.

This is an interesting account of the place of young people in residential care:

Edmond, Ruth (2003) Putting the care into residential care: the role of young people. *Journal of Social Work* 3(3) 321–37.

This article offers an interesting examination of the gendered discourses experienced by young Latinas in schools:

Hyams, Melissa (2000) 'Pay attention in class . . . [and] don't get pregnant': a discourse of academic success among adolescent Latinas. *Environment and Planning A* 32 635–54.

This book offers a rich insight into constructions of masculinity and sexuality in school spaces:

Mac an Ghaill, Mairtin (1997) *The making of men: masculinities, sexualities and schooling*. Buckingham: Open University Press.

This articles offers a useful overview of the ways in which geographers have explored the topic of institutions:

Philo, Chris and Parr, Hester (2000) Institutional geographies: introductory remarks. *Geoforum* 31 513–21.

This useful article explores the process of studentification and how it shapes the housing situations in cities and towns:

Smith, Darren and Holt, Louise (2007) Studentification and 'apprentice' gentrifiers within Britain's provincial towns and cities: extending the meaning of gentrification. *Environment and Planning A* 39 142–61.

10 Public space and the street

Richy: You can't go into a youth club at 17! Cos they're all young uns, aren't they? All there is is . . . it's a lack of everything. There's nothing to do, just streets to walk down and stuff like that. (MacDonald and Shildrick 2007: 343)

Susie: In the summer? Just go out hanging about.

Lara: Just walking around, walking up and down, looking, we walk about different roads. (Skelton 2000: 90)

Youthful publics

Richy hangs out in the street of his local area as he feels the youth clubs tends to cater for younger people rather than young people his age, whilst Susie and Lara talk about spending time in the summer hanging about and walking up and down the street. Clearly, public space and the street represent some of the most important places for young people in terms of how they construct and contest their identities, and how they create space for themselves as young people in an adultist society. Much research has focused upon the debates as to the meaning, use and understanding of public spaces and streets. Goheen (1998: 479) clarifies that public space in cities is:

> charged with meaning and with controversy. The space in question is that which the public collectively values – space to which it attributes symbolic significance and asserts claims. The values attaching to

> public space are those with which the generality of the citizenry endows it. Citizens create meaningful public space by expressing their attitudes, asserting their claims and using it for their own purposes. It thereby becomes a meaningful public resource. The process is a dynamic one, for meanings and uses are always liable to change. Renegotiation of understandings is ongoing; contention accompanies the process.

Furthermore, this perspective on public space rivals other approaches which focus on the importance of public space as the main site where all groups in society can 'achieve public visibility, seek recognition and make demands' (Goheen 1998: 480). Fyfe (1998: 1) also highlights the diversity of the ways in which streets can be imagined given that 'streets are the terrain of social encounters and political protest, sites of domination and resistance, places of pleasure and anxiety'.

Malone (2002: 163) highlights the importance of the street to young people as a 'stage for performance, where they construct their social identity in relation to their peers and other members of society':

> Many of the identities young people adopt within the public domain are contradictory and oppositional to the dominant culture (messy, dirty, loud, smoking, sexual); others have an easy fit (clean, neat, polite, in school uniform). Visible expressions of youth culture could be seen as the means of winning space from the dominant culture, to construct the self within the selfless sea of city streets; they are also an attempt to express and resolve symbolically the contradictions that they experience between cultural and ideological forces: between dominant ideologies, parent ideologies and the ideologies that arise from their own experiences of daily life.

As this quote demonstrates, young people are often regarded as deviant others when occupying public space and the street. Contrary to dominant discourses which represent children as being innocent and in need of protection within public spaces (Valentine 1996), there comes a point in the lifecourse when young people start to be identified as 'threatening presences' (Evans 2008: 1671) and as a threat to the moral order of the street (see Box 10.1). This transition – from innocent child to threatening other – is often regarded as happening at a younger age compared with previous generations. 'Excluded, positioned as intruders, young people's use of streets as spaces for expressing their own culture is misunderstood by many adults' (Malone 2002: 157).

Box 10.1

Folk devils and moral panics

Societies appear to be subject, every now and then, to periods of moral panic. A condition, episode, person or group of persons emerges to become defined as a threat to societal values and interests; its nature is presented in stylised and stereotypical fashion by the mass media; the moral barricades are manned by editors, bishops, politicians and right-thinking people; socially accredited experts pronounce their diagnoses and solutions; ways of coping are evolved or (more often) resorted to; the condition then disappears, submerges or deteriorates and becomes more visible. Sometimes the object of the panic is quite novel and at other times it is something which had been in existence long enough, but suddenly appears in the limelight. Sometimes the panic passes over and is forgotten, except in folklore and collective memory; at other times it has more serious and long-lasting repercussions and might produce such changes as those in legal and social policy or even in the ways that society conceives itself.

One of the most recurrent types of moral panic in Britain since the war has been associated with the emergence of various forms of youth culture (originally almost exclusively working class, but often recently middle class or student based) whose behaviour is deviant or delinquent. To a greater or lesser degree, these cultures have been associated with violence. The Teddy Boys, the Mods and Rockers, the Hells Angels, the Skinheads and the Hippies have all been phenomena of this kind. There have been parallel reactions to the drug problem, student militancy, political demonstrations, football hooliganism, vandalism of various kinds and crime and violence in general. But such groups as the Teddy Boys and the Mods and Rockers have been distinctive in being identified not just in terms of particular events (such as demonstrations) or particular disapproved forms of behaviour (such as drug-taking or violence) but as distinguishable social types. In the gallery of types that society erects to show its members which roles should be avoided and which should be emulated, these groups have occupied a constant position as folk devils: reminders of what we should not be. The identities of such social types are public property and these particular adolescent groups have symbolised – both in what they were and how they were reacted to – much of the social change which has taken place in Britain over the last twenty years.

Adapted from Cohen (1987: 9–10)

This series of misunderstandings leads to problematic assumptions and stereotypes being reinforced about young people's use of public space and the street, creating moral panics about young people's presence in public space and reinforcing mythologies about young people:

> [M]ythology is defined as an often vague and diffuse way of imagining particular 'real' places and the people in them. Such myths may be considered as a set of 'stories' about a place, stories whose origins and characteristics are difficult to pin down but become widely known, and often accepted, as having some basis in truth. Inevitably, these stories or myths serve to stereotype a particular place or set of places by highlighting some of its characteristics in favour of others
>
> (Holloway and Hubbard 2001: 117)

The idea of public space raises concerns, especially where young people are concerned, such as those relating to 'the panic over "wilding" in New York's Central Park in the late 1980s (rampaging young men violently terrorizing joggers and other park users for the sheer joy of it)' (Mitchell 2003: 13).

As Malone (2002: 162) points out, 'the visibility of youth and their competing use of street spaces positions them in the front line of conflict over its use', yet a key issue is the lack of consideration given to young people's perspectives about public space, the diversity of their experiences and the agency they often exercise in their negotiations of public space and the street. For many young people, there is a lack of community facilities available to them, and so the public spaces and streets of their neighbourhoods offer them some of the only places where they can spend time. Matthews *et al.* (1999b: 1716) talk about the street as a social arena for young people and they note that 'contrary to media stereotypes, the street is no longer a male-dominated terrain.' The street is often regarded as a place of fear for young people when alone, but much less so when with friends (Matthews *et al.* 1999b) and so the presence of others makes young people feel more secure

Young people's engagements with the street and public space vary by age and time. Older young people have tended to be afforded greater spatial freedom than younger children and tend to be regarded negatively compared with children who are associated with innocence and regarded as being in need of protection (Valentine 1996). As Matthews (2003: 104) found in his research with teenagers in England,

> the ways in which young people used the street varied by age. For those aged 11 and under, the street was a setting for games, play and adventure. By the age of 13, however, the street was a social haven, a place for meeting with friends, hanging out and 'where things

> happened' . . . for older teenagers, in particular, the street offered opportunities 'to get away from it', sites that offered the freedom and excitement of separation away from the 'humdrum' of daily life.

Similarly, MacDonald and Shildrick (2007: 342) found that for working-class young people in northern England, when they were not at school or in part-time jobs, the majority of their time was spent with friends in 'the public spaces of their home estates.' They note that the gendered separation of men as public and women as private no longer exists. They also detected that although young people may avoid youth clubs due to the presence of younger children, there is a point where they may start to engage with adult forms of leisure and consumption, such as those focused on alcohol-consumption. This move from the street to the spaces of the pubs and clubs often related to reaching the minimum school-leaving age. Young people spent less time hanging out in the public spaces and streets of their neighbourhood to focus their energy on the pubs and clubs of the town centre. Clearly then, young people's use of the street and public space alters as they negotiate their journeys through the lifecourse.

Another important theme in work about young people, the street and public space is the way in which they demonstrate their agency and creativity in their use of space:

> In order to remain on the street, therefore, young people carve out their own cultural crevices, and create their own social fissures. Often these are places where adults are not commonly found: in this study children regularly congregated in back alleys, on derelict land, around lock-up garages, at the rear of shopping parades, in pockets of green space within neighbourhood scrub woodland, in essence, within the forgotten and redundant spaces of the adult world.
>
> (Matthews 2003: 106)

This led Matthews (2003) to observe a number of 'special places' of teenagers, many of which were unknown to adults and beyond their immediate surveillance. According to Matthews *et al.* (1998: 195), young teenagers' spatial experiences are different from what adults often imagine:

> [C]hildren's play areas became convenient places where groups could hang out during the evening away from the adult gaze; the local shops became a social venue where teenagers from one group could come into contact with other groups and show-off their latest clothes and

> hairstyles, and wait for things to happen; and alleyways and back passages provided spaces for exciting mountain bike races.

This all took place in the context of persistent attempts to control the spatial movements or behaviours of young people. Overall then, 'streets are places where adultist conventions and moralities about what it is to be a child – that is, less-than-adult, can be put aside. They are spaces that are temporarily outside adult society, particularly with the withdrawal of adults at particular times of the day' (Matthews 2003: 106). Matthews (2003) also notes that the street is important for young people, especially during the summer months and after school, with more than two-thirds of young people spending over six hours outside. This provides one of the only places for some young people to meet informally (see Box 10.2).

Furthermore, as Cahill (2000: 251) observes, many teenagers 'have a highly developed understanding of environmental protocol and can "read" the environment in specific ways that are at once personal, cultural and social.' Cahill develops the idea of 'street literacy' which is 'an interpretative framework that privileges experienced informal local knowledges that are grounded in personal experiences and passed down in the form of rules, boundaries set by parents, neighbourhood folklore, and kids' collective wisdom.' She argues that young people have a deep and subtle appreciation of their local streets evidenced by their daily negotiations and knowledges of their local environments. Furthermore, in the everyday reclaiming of space, 'young people may leave their own territorial markers as symbolic gestures of their distancing from the work of adults' (Matthews *et al.* 1998: 197).

Box 10.2

The 'special places' of teenagers

Places away from authority	Outdoors, woods, fields, streets, back lanes
Places to be with friends	Woods, parks, play areas, streets, local shops, sports centres, village green, town centre shopping mall, friend's house, own house
Places for adventure	Woods or forests, local lakes, streets, back alleys, underpasses, building sites, derelict land
Places for solitude	Woods, bedrooms, backyards, garden

Adapted from Matthews *et al.* (1998: 198)

Image 10.1 Young people in Amsterdam.

In this chapter, I focus on hanging out, curfews, skateboarding and young people making public space in order to explore the complexity of the street and public space for understanding the intersection of young people, place and identity (Image 10.1). An important point to consider here is the way in which these discussions tend to be adult-focused with a lack of attempt to appreciate young people's perspectives on the street and public space (MacDonald and Shildrick 2007).

Hanging out

As mentioned above, many young people choose to use public spaces and the street as one of the few places where they can socialise with friends in a context that is away from the institutional restrictions of school or the watchful eye of family members. Matthews *et al.* (2000) observe that there is a lack of places for young people to hang out. Whilst adults are able to withdraw to the confines of their homes, workplaces or pubs, young people tend to be restricted to particular public places, such as the street or local shopping malls. The vast majority of young people in the research of Matthews *et al.* (2002) visited a shopping mall weekly with some visiting every day. Daily visits also increased with age. Young people liked the shopping mall because it was warm and dry and they could hang out and meet up with friends. There was also a degree of social credibility associated with hanging out with friends in the mall. However, the mall is also a site of conflict as young people were regarded as occupying a traditional adult space as well as not participating in the consumerism expected to take place there. Half of the young people had been asked to move on, yet many felt that such requests were very unfair and unjust. Some moved to other parts of the mall, with others leaving and returning after a short while, with a very small proportion not returning at all.

For many young people, hanging out with friends – whether it is in a shopping mall or in other public spaces – is a crucial part of their everyday lives. Their friendship groups 'provide a range of emotional and social support and are also major sources of knowledge and understanding' (Hill *et al.* 2007: 17). Furthermore, 'many studies highlight the protective aspect of peer relationships' (Hill *et al.* 2007: 16) and they can have an important role to play in terms of young people's health, well-being, social skills, schooling and overall behaviour. Drawing upon research with young women in the USA, Thomas (2005: 591) sees 'hanging out' as 'a broad term that encompasses a range of social activities' including 'walking or driving around, shopping, sitting and talking in public or private space, lounging, and watching TV.' This positive definition of hanging out is important as there is a tendency for young people in groups to be demonised and seen as a threat. Hill *et al.* (2007: 17) note that there 'is a tendency for adults to see peer influences as negative and so demonise peer associations'.

Despite the reasonably positive associations of the phrase 'hanging out', a group of young people occupying public spaces tends to be identified as a 'gang', a term with far more negative consequences. As Collins *et al.* (2000: 137) observe:

> The gang has a long history in social analysis – it is, however, a problematic category indicative more of the social anxieties around the perceived deviancy of marginalised groups than of adolescent group formation . . . on the one hand it is used to identify forms of criminal association, ranging from pulp fiction evocations of frontier banditry in nineteenth-century United States, to modern forms of

professional crime networks (such as Asian Triads). On the other, it is used to describe forms of adolescent or childhood association in leisure activities.

Matthews *et al.* (1998: 196) observed in their work that despite the gangs of young people being 'unnamed, each was identifiable by their territory, patterns of behaviour and their 'style'. Importantly, these gangs were seen as groups of friends who liked to participate in similar pastimes and would often spend time hanging out together in parks, vacant land or other spaces available to them. Some of these groups formed around shared interests in sport, others were more interested in fashion. Here, Matthews *et al.* (1998: 196) find the concept of 'microcultures' useful and state that these are 'created by combinations of personalities, the locations they make their own and the events they share. Together these provide a common and unifying set of experiences.' They critique the 'youth subcultures' approach to understanding young people's lives, arguing that a focus upon 'microcultures' appreciates the diversities and differences within young people's lives, rather than seeing them as being apart from wider society.

Dominant discourses of 'the gang' tend to focus upon the territorial and group behaviour of young people, often drawing attention to the behaviour, attitudes and values of young men from working-class or minority ethnic backgrounds. The group practices of working-class white youth have been the focus of much research about youth subcultures (e.g. Willis 1977), with relatively less attention being given to black youth, and Asian youth culture being almost ignored completely. Yet, there has been a remarkable change in focus in recent years. As highlighted in the work of Alexander (2004: 531) the focus of discussions about social breakdown, cultural marginality and dysfunction is often upon 'Muslim young men, encapsulated in the image of "the gang". There is a clear "picture of angry young men, alienated from society and their own communities, entangled in a life of crime and violence".' The idea of 'the gang' draws strongly upon everyday discourses which see Asian masculinities as 'collectively dysfunctional and as newly dangerous' (Alexander 2004: 532). Where Asian young men were previously largely ignored by academic research as a result of assumptions about their inheritance of 'patriarchal privilege', 'they are now the hyper-visible embodiments of a racialized dysfunctionality' (Macey 1999). This is signalled most clearly in the shift from 'victim' to 'aggressor' status (Alexander 2004: 535).

The behaviour of young people in public space, where they hang out and what they spend their time doing is often a concern for people who live and work close to where young people choose to hang out. One response to this issue has been the development of the Mosquito which is a 'device that is designed to "repel" young people through the emission of a high-pitched sound set at between 18 and 20 kilohertz' (Walsh 2008: 122). The noise emitted from a Mosquito is often at the top end of the hearing range of most teenagers whereas younger and older people may not hear it at all or may only hear it slightly. Charlotte Walsh (2008) notes that over 3,000 Mosquitos have been sold to private purchasers and public bodies across the world. She explores the use of the Mosquito in terms of the law and argues that the UK government should intervene given that the use of the device is not acceptable in terms of human rights legislation. She also questions the use of the Mosquito in the context of environmental regulations alongside the ways in which its use is antisocial and a form of criminal harassment. Walsh (2008: 132) highlights the creative response of young people to the use of such a device when she notes:

> It is the teenagers themselves who add a final ironic twist . . . through their appropriation of the Mosquito's buzz as a ring tone: playing with the fact that adult teachers cannot generally hear it, this is the perfect text-message alert system to use in the classroom.

Imposing curfews

Debates about the use of streets and public spaces by young people often result in discussions about the ways in which their engagement with such places can be monitored, controlled and limited. Curfews, which place limitations upon when young people can use public spaces,are often presented as the answer to concerns about youth crime, gang activities and the 'problem' of young people in public space. For Collins and Kearns (2001: 389) the use of curfews is symbolic of the adult control of public space:

> Public space is being constructed as adult space through legal mechanisms such as curfews, which seek to curtail young people's spatial freedom and contain them within their homes. Ostensibly motivated by a desire to reduce youth crime and victimisation, curfews reflect a contemporary preoccupation with achieving social control through the control of space.

Likewise, Matthews *et al.* (1999: 1715) have observed that 'the curfew debate is embedded in "adultist" constructions which relate to the spatial ordering of society by age and in assumptions which fail to acknowledge young people as active social agents'. This results in young people experiencing strong senses of disconnection from society as their movements are constantly limited and restricted.

Drawing upon work in Perth, Australia, Iveson (2007: 171) discusses the 'Young People in Northbridge Policy' which essentially

> instructs police to use their powers under Child Welfare legislation to remove unaccompanied children under 12 from the inner city 'adult entertainment precinct' of Northridge after dark, to remove children from 13–15 after 10pm, and to subject all others aged under 18 to a more 'hard line' approach. In the first year of the curfew's operation, 961 young people were removed from the streets in Northbridge. A staggering 88 per cent of those young people were Aboriginal.

As highlighted here, aside from the problematic nature of curfews in general, they can result in young people with particular identities becoming the victims of policy as racialised and ethnicised youth are targeted due to problematic stereotypes about their behaviour in public places. Furthermore, Iveson (2007) also points out that one of the criticisms of curfew policy is that there is a need to offer 'youth-friendly' spaces rather than clearing young people from the street.

Another example of curfew policy is highlighted by O'Dougherty's (2006) discussion of the Mall of America, Minnesota, USA, where curfews were used as a mechanism for controlling young people's use of space. Implemented in 1996, the rules of the curfew state that young people under the age of 16 must be accompanied by an adult of at least 21 years of age after 6pm on Friday and Saturday evenings. Walsh (2002) sees the use of curfews as emanating from moral panics about the place of young people in society coupled with government aims to impose their views about what young people should be doing onto young people. Curfews clearly work to support the domination of adults in society, as young people are persistently demonised, criminalised and restricted in their use of particular places (Collins and Kearns 2001: 393).

Matthews *et al.* (1999) have identified ten arguments against the use of curfews. They question the legality of imposing curfews on young people in the UK, arguing that similar restrictions on adults would never be permitted and that placing such limitations on young people breaches the European Convention on Human Rights. Furthermore, given the value young people place on the street as important for socialising, preventing them from using such places greatly limits their opportunities for hanging out with friends and gaining social skills:

> Curfew as a solution to (adult) society's concerns about the behaviour of some, challenges the notion of young people as responsible citizens and both potentially taints the majority because of the actions of a few and punishes the innocent for the antisocial behaviour of a minority.
>
> (Matthews *et al.* 1999: 1722)

Moreover, imposing curfews is also likely to discriminate against the least affluent who already have restricted opportunities compared with their richer counterparts who have greater social mobility. As such, the imposition of curfews is likely to have quite different influences on young people depending on where they are and how their social identities are interpreted.

A curfew on young people makes an assumption that age, rather than gender, is the major factor in influencing crime and will work to further marginalise young women, a group who are already regarded as being excluded within public space. Curfews also 'mask the plight of those young people who are staying out for particular reasons, such as to keep away from drunken, violent or abusive parents' (Matthews *et al.* 1999: 1723). Given that the places most likely to impose curfews are those on the margins of society, imposing further restrictions is likely to stigmatise and isolate them even further. The place-based effects of curfews are also likely to reshape the experiences of other places as young people may choose to hang out in neighbouring communities where curfews are not in place. Using a curfew also does not get to the heart of issues such as poor parenting, familial breakdown or antisocial behaviour and instead acts to conceal the issue. There is also a striking lack of evidence of such approaches being effective. One of the most important criticisms is the lack of involvement of young people in processes and decisions about curfew policy:

> Young people are seemingly invisible on the landscape, in that most physical environments are built by adults, for adults, with scant understanding of young people's needs and aspirations.
>
> (Matthews *et al.* 1999: 1725)

Skateboarding

According to Woolley and Johns (2001: 212), it is during the teenage years that 'young people further establish their self-identify through opinions, values, looks and preferences, choosing musical styles, dress codes and leisure activities as the building blocks of self-identity.' Focusing upon young skateboarders, they note that skateboarders tend to prefer hard and urban locations for their pursuits, and they tend to occupy such spaces in very creative and exciting ways. Skateboarding locations 'vary from university campuses, office plazas and urban squares to the more everyday spaces of streets, pavements and car parks' (Woolley and Johns 2001: 214). A key theme in research about young people and skateboarding is the tension between skaters' desires to practise their skills and socialise with their peer group set against the ways in which they are often regarded as problematic users of public spaces by other users, including the local authorities as well as other adult users of the space. Skaters are often regarded as jeopardising the aims of city managers and officials and are often seen as a 'problem' and stereotyped as out-of-control, rebellious and uncooperative 'youth'.

As one skateboarder said in research by Woolley and Johns (2001: 226):

> People don't understand us – they think we're vandals – we don't go out of our way to damage stuff, we're just using it . . . we enjoy using the stuff that nobody else even notices. What's wrong with that?

Stratford (2002) conducted research into the connections between skating and urban governance, focusing in particular upon Franklin Square in Hobart, Tasmania, Australia. At certain times of day, skaters tended to be the dominant users of the Square, in particular before and after school as they waited for buses or were hanging out with friends. Yet, for the young people, there was a strong sense of always being under surveillance. Related to this, Owens (2002: 158) discusses communities in the USA that have adopted 'ordinances restricting the use of skateboards in defined public areas'.

Karsten and Pel (2000) found that most of the skateboarders in Amsterdam were in their twenties; as younger children they tended to practise on local streets with the older young people venturing out into the city to practise their skills. They found that 96 per cent of skaters in Amsterdam were boys of mainly white middle-class backgrounds. For many, it was more than just sport or a form of leisure and was instead an 'identity-building performance' (Karsten and Pel 2000: 338). They noted that 'the way in which skaters use the city is essentially different from that of the rest of its inhabitants and visitors' (Karsten and Pel 2000: 327). Young men would gather together in skate shops or in skate parks to repair their skateboards, share tips and hear about the latest spots for skating, and so it is 'an individual sport as well as a social event' (Karsten and Pel 2000: 335). Important issues for young skaters tended to focus on accessibility, trickability, sociability and compatibility of the local landscape, along with musical preferences and dress choice (Borden 2001).

Borden (2001: 137) notes that the 'identity of skateboarders continually informs, and is informed by, their spatial activities' and queries the extent to which skateboarding can be associated with particular performances of gender and sexuality. He questions the extent to which skateboarding could be regarded as a homosexual activity, given that skateboarding tends to involve young men spending lots of time watching each other, performing in front of each other, emulating other skateboarders as well as reading magazines and looking at images of other skateboarders. An alternative reading could be that skateboarding acts to bolster the heterosexuality of the young men participating in it, just as many other sports do. However, skateboarding could also be regarded as an individual practice and so the key element may be self-satisfaction and self-improvement as individual moves and styles are practised and improved. The baggy clothes worn by some skaters – especially in the 1980s – may help flexibility and give skaters more freedom to perform particular moves; however, this could also be seen as a de-sexualising of the body, as the shape, form, masculinity and other markings of the body are hidden from view (Borden 2001, Karsten and Pel 2000).

Making public space

The overwhelming focus of discussion about young people and public space focuses concerns and anxieties about the ways in which young people present a threat to the moral order of civilised society. Much of this points to ways in which young people's presence in public space can be controlled, limited and restricted in particular ways. Despite this negative focus, there now exist numerous examples of young people creatively making public space in ways which challenge, overlook and pass by adult assumptions and regulations about how streets and public spaces should be used. One such example can be found in recent work about parkour (see Box 10.3) a physical activity which focuses upon moving from place to place as quickly and as efficiently as possible (Saville 2008). In this final section, I focus upon the ways in which young people use cars and mobile phones in ways that assist them in making use of public space and gaining autonomy away from the stereotypes, demonisation and control of adult society).

Falconer and Kingham (2007) discuss the experiences of 'boy racers' in Christchurch, New Zealand. 'It was clear to us that those driving the "flashier" cars . . . were at the top of the social hierarchy' (Falconer and Kingham 2007: 15). Their cars received the most attention and were the ones that other participants most wanted to ride in. Despite the label 'boy racer', they found that both young men and women participated in this engagement with public space. Although some young women were identified as hanging onto young men who drove the best cars, there were a number of young women who would race young men in order to get noticed or to challenge the gender order. There was a variety of reasons why young people chose to participate: to gain respect and recognition from friends, to show off to friends, to socialise, to get away from the stresses of everyday life, to experience a rush, to challenge the law and to avoid being bored. Some young people were not necessarily interested in cars nor did they own a car,

Box 10.3

Parkour

On Wednesday 17 May 2006, David Belle, a man heralded as a founder of parkour, and now international celebrity, runs towards a solid wall that cordons off an underpass. His movement is purposeful and practised. All eyes are turned on him. As he closes on the barrier and jumps, his calloused hands reach out, pushing off the tops to propel him over. His trailing foot does not clear the wall; it hits. It pulls his body out of alignment. Travelling at speed and out of control David Belle's back makes contact with the far wall of the underpass and he falls . . .' (Saville 2008: 891)

Parkour is often described as free-running and is essentially a physical activity the aim of which is to get from one place to another as quickly and efficiently as possible by overcoming boundaries and limitations within the environment. Practised by traceurs (parkour practitioners), parkour is a diverse set of practices drawing upon a range of physical movements such as climbing, jumping or vaulting.

but they participated in this form of leisure activity because they wanted to spend time with their friends. Although the ownership of a good car and a good performance improve social positioning, this was not the only intention of 'boy racers'. Other closely aligned activities included alcohol consumption, stunt driving, damage to property, selling and using drugs and engagement in violent and threatening behaviour. The young people were able to relocate to other locations if their activities came under the gaze of restrictive or regulatory bodies or individuals. Such practices tended to be concentrated during the weekends when young people have more time. One young man, 'Constable Keith' said:

> Bro, if I didn't have my own car, I'd hang out with [associate] . . . His car is sweet [a late-model Nissan Pulsar with plenty of add-ons] and he gets all the girls wanting a ride.

Another example of young people's engagement with car culture is demonstrated in Vaaranen and Wieloch's (2002) ethnographic work about working-class young men in Helsinki, Finland. Although they found that young men in particular often engaged in driving enthusiastically, embracing the music, speed and associated risks, they also argue that such practices worked to reinforce their class position. As these young people developed their knowledge and sense of control over driving, music and speed, the bonds they formed connected them into their social class grouping.

Also focusing upon young people, cars and driving, Redshaw and Noble (2006) conducted ten focus groups with 65 young people across urban and rural New South Wales, Australia, about the complexities of driving practices. One of the most common ways of framing the importance of having a car is the instrumental rationalisation of 'getting from A to B' (Redshaw and Noble 2006: 10). However, more than this, having a car enabled young people to travel further afield and offered them a sense of empowerment and control given their greater mobility and freedom. Driving offered them greater access to the key domains of young people's lives – work, education, leisure, consumption sites – and was particularly attractive for rural or suburban youth who, through having a car, were more able to access more desirable locations in the city. As Redshaw and Noble (2006: 13) note 'the mobilities that come with having access to a car enable young people to do things and go places that otherwise would be difficult or unthinkable.'

Having a car offers young people more independence from friends and family alongside increased opportunities for spontaneity. However, by driving, young people are required to restrict their intake of alcoholic drinks. Having to rely on others when their car is being repaired starkly highlights what the car offers to young people's freedom. Redshaw and Noble (2006) note that men outnumber women in most forms of road trauma, including as cyclists, pedestrians and drivers, with women outnumbering men only as passengers. Furthermore, young drivers actively produce their gendered identities through driving and through their ownership of particular types of car. Young people argued that there were 'girl cars' and 'guy cars' which were often differentiated in terms of size, style and performance as well as how they were used, who drove them and whether or not they had been modified in any way. Some of the young people naturalised gender differences by arguing that because men are physically bigger, they need bigger cars, yet for some men, spending time on their cars as well as cruising around were key practices that worked to bolster their masculine identities.

Alongside cars, mobiles, or cell phones, are often regarded as highly desirable accessories by young people. Furthermore, different ring tones, colours and covers mean that they can be seen as a mechanism whereby young people articulate their identities. It has been observed that 'relatively little attention has been given to the role of mobile phones in the management of young people's safety, or in influencing their movements in time and space more broadly' (Pain *et al.* 2005: 815). Arguably mobile phones have 'opened up new spaces for social interaction; for friends to talk, text and arrange to meet, and for parents and young people to stay in touch' (Pain *et al.* 2005: 815). Potentially, mobile phones may increase young people's geographical freedom, as parents feel comfortable allowing them to spend time hanging out with friends, knowing that they are only a phone call away. Parents may also feel that there is less need to accompany their children and so young people may have a wider unsupervised spatial range. This also means that there is scope to restrict the information that young people provide to their parents and so they may not tell the truth about where they are. Technically, however, mobile phones can also be used to monitor where children are. However, it is important to recognise that not all young people have access to a mobile phone.

Although mobile phones offer new autonomous spaces and freedom away from the home, their use is bound up with many factors relating to the social position of young people, including their relationship with their parents or carer; whether they receive pocket money and/or have an income from working outside the home; where they live; whether their friends live close by and whether these friends have a mobile (Pain *et al.* 2005: 826). Furthermore, although mobile phones may offer greater spatial freedom to young people, they are also 'opening up new spaces of risk' (Pain *et al.* 2005: 816) as young people may be bullied through the use of mobile phone technology, technology that is not easily supervised by parents. Besides, some young people find that they cannot afford a mobile phone and so have restricted access to such commodities. So, although mobile phones may be decreasing concerns about crime and risks in society, they also open up new spaces of victimisation, and so are reshaping rather than reducing parents' concerns about young people in public spaces.

These concerns about bullying have led some schools in New Zealand to ban mobile phones. Research by Thompson and Cupples (2008) in New Zealand highlights that texting is important for young people as a mechanism for engineering face-to-face meetings and for building up connections with friends and expanding their social networks. Text messages are constructed as something private, unseen and unheard and are often preferred over telephone conversations as they are generally cheaper. Thompson and Cupples (2008: 100) argue that 'cell phones are commonly used in ways which escape adult surveillance'. Cell phones 'facilitate the development of new socio-spatial relationships which provide young people with creative and interactive modes of negotiating public and private space, the body, surveillance and authority' (Thompson and Cupples 2008: 104).

Key themes

- Popular understandings of young people often reinforce problematic stereotypes about their use of the street and public spaces, yet there is much evidence that public spaces are some of the only sites available for young people to spend time outside the home or adult-dominated institutions.
- Research about young people and public space often focuses on their experiences of hanging out in the street, in shopping malls or in other public spaces, and emphasises the ways in which the adult domination of public places work to monitor young people's behaviours.

- The imposition of curfews is one example of the ways in which adults seek to maintain control of the street and public space, yet the use of curfews demonstrates a number of problematic assumptions about young people, place and identity.
- Although there is much emphasis on the control and surveillance of young people who occupy the street and public space, research has demonstrated the ways in which young people make public space in creative ways through, for example, parkour, driving cultures or the strategic use of mobile phones.

Project ideas

Find a selection of newspaper articles about young people's use of public space and explore the dominant representations of age, place and identity in them.

Select a public space in the city or town where you live and explore the ways in which young people use this space. What particular groups of young people occupy this place and why? How does their use of this space change through time? What processes of control or surveillance are in operation and how do young people respond to such mechanisms?

What mechanisms, processes and policies are in operation to control the ways in which young people use public space and the street?

Explore the ways in which young people use public space and the street to challenge adult-centred uses of place.

Suggested further reading

This book gives a critical account of skateboarding in the context of the city and the body:

Borden, Iain (2001) *Skateboarding, space and the city: architecture and the body*. Oxford: Berg.

An important contribution to youth studies, this text explores the ideas of folk devils and moral panics:

Cohen, Stanley (1987) *Folk devils and moral panics: the creation of the mods and rockers*. Oxford: Blackwell.

This introductory chapter provides a useful overview of understandings of the street:

Fyfe, Nicholas R. (1998) Introduction: reading the street. Fyfe, Nicholas R. (ed.) *Images of the street: planning, identity and control in public space*. London: Routledge. 1–12.

This excellent article offers a series of insights into the multiple understandings of public space and the city:

Goheen, Peter G. (1998) Public space and the geography of the modern city. *Progress in Human Geography* 22(4) 479–96.

This article offers an important series of arguments about the use of curfews in controlling public space:

Matthews, Hugh, Limb, Melanie and Taylor, Mark (1999) Reclaiming the street: the discourse of curfew. *Environment and Planning A* 31 1713–30.

This interesting article charts a range of issues associated with skateboarding and how it changes the meanings of the city:

Woolley, Helen and Johns, Ralph (2001) Skateboarding: the city as a playground. *Journal of Urban Design* 6(2) 211–30.

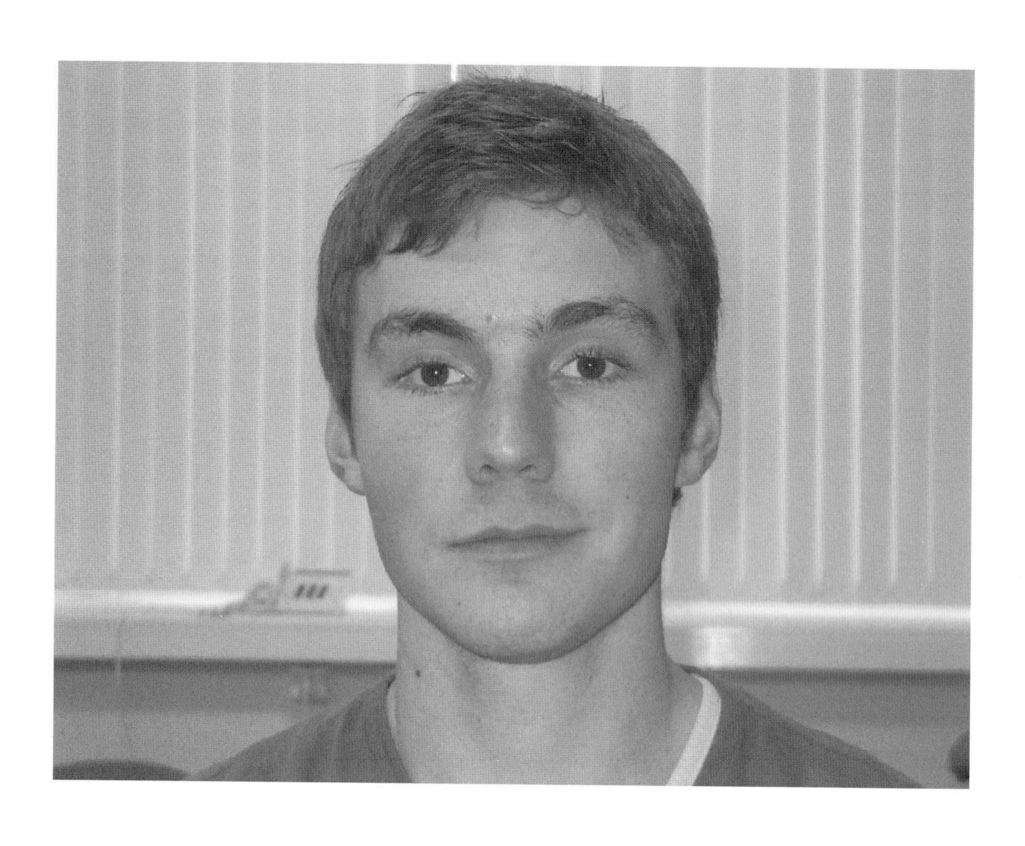

11 Migrations, mobilities and transitions

Lisa: Sometimes I feel like I've reached adulthood, and then I'll sit down and eat ice cream directly from the box, and I keep thinking, 'I'll know I'm an adult when I don't eat ice cream right out of the box anymore' . . . But I guess in some ways I feel like I'm an adult. I'm a pretty responsible person. I mean, if I say I'm going to do something, I do it. Financially, I'm fairly responsible with my money, But there are still times where I think, 'I can't believe I'm 25.' A lot of times I don't really feel like an adult. (Arnett 2001: 17)

Linda: Well, when I first arrived people couldn't understand what I was saying and I couldn't understand what they were saying because they were talking so fast. (Hopkins and Hill 2006: 44)

Youthful migrations, mobilities and transitions

In the quotations above, Lisa is reflecting upon her transition towards adulthood with particular reference to her financial and associated responsibilities, whilst Linda talks about her experiences of arriving in Scotland as an unaccompanied asylum-seeking young person who had fled from persecution in her country of origin. Lisa and Linda are – in very different ways – reflecting upon the ways in which experiences of migration, mobility and transition intersect with their identities and practices as young people. It is clear that issues of migration and mobility are central to how young people live their lives. Whether it be about the journey to school, leaving home for university, taking a gap year or moving out of the family home, it is clear that issues of migration and transition are fundamental to what it means to be young. Yet, young people's experiences of migration and mobility are relatively under-researched. As Barker *et al.* (2009: 2) observe, 'despite interest both within the new mobilities paradigm and children's geographies, only a relatively small number of articles consider debates regarding children's and young people's mobility'. Yet, as Sheller and Urry (2006: 207) note:

> All the world seems to be on the move. Asylum seekers, international students, terrorists, members of diasporas, holidaymakers, business people, sports stars, refugees, backpackers, commuters, the early retired, young mobile professionals, prostitutes, armed forces – these and many others fill the world's airports, buses, ships, and trains. The scale of this travelling is immense. Internationally there are over 700 million legal passenger arrivals each year (compared with 25 million in 1950) with a predicted 1 billion by 2010; there are 4 million air passengers each day; 31 million refugees are displaced from their homes; and there is one car for every 8.6 people.

One of the main reasons why there is a lack of work about young people, migration and mobility is due to the fact that population geographers and demographers have tended only to study adults or family units. As John McKendrick (2001: 466) notes, 'it is clear that children feature prominently throughout population geography. However, the population geography of childhood is a mirage, in that children are ever-present, but never really there'. For McKendrick, the voices of children and young people are not heard in population geography and demography, yet it is clearly important to see children and young people not only as subject to various forms of migration and mobility but as having agency in influencing, shaping and moulding such experiences.

Barker *et al.* (2009) note that for young people – particularly for children under the age of 10 – levels of independent mobility are in decline and that cars are becoming a frequent means of travel for young people. Research has also shown that children from working-class families are likely to have less space available to them in the family home and so are often likely to spend time unsupervised outdoors. For young people from upper-class backgrounds, this is often not the case as they may experience restricted independent mobility and tend to be escorted by adults in a car. Furthermore, the gendering of young people's experiences suggests that boys and young men are afforded greater spatial freedom and enhanced opportunities for mobility, although recent work suggests that this might no longer be the case (Barker *et al.* 2009). As such, young people's experiences of migration, mobility and transition are strongly shaped by social identities such as class and gender.

An important consideration about young people, migration and mobility is to recognise that ways in which age and mobility are simultaneously produced in different ways. As Barker *et al.* (2009: 6) observe, 'flexible, youthful identities privilege certain kinds of mobilities, encompassed by opportunities as diverse as international travel, global communication, education status and personal financial power.' However, they also note that 'social studies of mobilities must – always, already – attend to the importance of age, ageing, and lines of aged difference which are part-and-parcel of so many mobilities' (Barker *et al.* 2009: 7).

An increasingly significant form of migration relates to young people who move to study at university yet 'the standard academic literature on migration pays virtually no attention to students as migrants' (King and Ruiz-Gelices 2003: 230). Students who have the economic resources to move for study can be regarded as an elite group, compared with those who are restricted to local universities or colleges. From work in a number of European countries, it was found that students were more likely to engage in international mobility to study abroad if this was part of their pre-university experiences or if they already had knowledge of foreign languages. King and Ruiz-Gelices (2003) draw upon data from UNESCO when noting that there were 1.6 million tertiary-level students studying abroad in 1996 (1.35 million five years previously) – half of these were studying in Europe, 37 per cent in North America and 7 per cent in Australia. Europe accounts for a third of all sending countries. They also observed the influence that EU-financed schemes such as Erasmus and Socrates have on international student mobility noting that in 1999–2000, the UK received 20,705 students and sent 10,056.

British students' motivations for a year abroad include the cultural experiences of living in another country, improving their language proficiency as well as working on their professional development. Year-abroad students are also more likely to engage in postgraduate study and have higher salaries, better job profiles and fewer accounts of unemployment compared with students who do not take a year abroad. For many of these students, the experience of having a year abroad in Europe heightens their sense of connection and attachment with a European identity, and many regularly visit the country they studied in or even choose to live there for part or all of their lives. These connections, although shaped by the year-abroad experience, may also have been influenced by visits to Europe at earlier stages of their lives. Waters (2006a, b) has conducted research with immigrant students in Vancouver and returnee graduates in Hong Kong who had completed their degrees in Canada. The immigrant students studying in Canada all stressed the importance of family expectations and goals and over a half planned to return to Asia. This work emphasises the spatial strategies adopted by upper- and middle-class families in their attempts to bypass a supposedly failing local education scene (Waters 2006a).

Students moving for study provide one example of the ways in which migration, mobility and transition are key experiences of many young people today, and there are many other examples. In this chapter, I focus upon three very different examples of migration, mobility and transition in young people's lives in order to explore the ways in which young people, place and identity shape and are shaped by experiences and accounts of migration and mobility. First, I discuss the experiences of young people who have been forced to move from their countries of origin due to experiences of persecution. Second, I explore literature about the transition to adulthood, and finally I discuss young people's engagement with the gap year.

Fleeing persecution

According to the 1951 Convention on the Status of Refugees, the term 'refugee' shall apply to any person who,

> owing to well-founded fear of being persecuted for reason of race, religion, nationality, membership of a particular social group or political opinion, is outside the country of his nationality and is unable or, owing to such fear, is unwilling to avail himself of the protection of that country; or who, not having a nationality and being outside the country of his former habitual residence as a result of such events, is unable or, owing to such fear, is unwilling to return to it.

As such, a refugee is a person who has fled persecution and has found safety in another country having successfully applied to stay in that country. An asylum seeker is a person who has fled persecution in their country of origin, has identified themselves to the relevant government or associated authorities and is exercising their right to apply for refugee status under the 1951 Convention. It is important to distinguish the terms 'refugee' and 'asylum seeker' from the term 'economic migrant', as the first two are associated with protection issues, and economic migrants are associated with moving in order to work. According to the United Nations High Commissioner for Refugees (2008a), there were 11.4 million refugees in the world, with the vast majority of refugees seeking asylum in countries neighbouring their country of origin. In 2007, Pakistan, Syria, Iran, Germany and Jordan were the five most popular destinations for those seeking refuge, and Afghanistan, Iraq, Somalia, Burundi and the Democratic Republic of Congo were the five most frequently mentioned countries of origin of refugees. Twenty-seven per cent of the global refugee population are from Afghanistan.

Often children and young people who are refugees or asylum seekers apply for refugee status as part of a family. However, some may seek asylum with other relatives or family members and some also seek asylum alone (Box 11.1). Many young people who are refugees were previously unaware of what the label 'refugee' means; yet they are essentially provided with a new form of identity which they may or may not associate with in very positive ways. As young refugees, their identities are embodied in terms of their skin colour, accent or clothing, yet their identities are also bound up in connections with issues of the nation-state. As Jo Boyden and Jason Hart (2007) note, young people who are displaced by armed conflict and political oppression are connected with issues of the nation-state due to the fact that those associated with the state are often the main perpetrators of violence against citizens.

In research conducted with unaccompanied asylum-seeking children and young people in Scotland, Hopkins and Hill (2008) explored the pre-flight experiences and migration stories of the young people involved. Through interviews with service providers as well as group work and individual interviews with separated children, they identified a range of pre-flight experiences. These included personal persecution or the death or persecution of family members; personal experiences of war and violence; and oppression and aggression from government as a result of religious of political affiliations. Clearly then, the ways in which young people (and their families) identify with – or are identified by others as being associated

Box 11.1

Separated children

'Unaccompanied asylum seeking children and young people' are often referred to as 'unaccompanied children', 'unaccompanied minors', 'separated children' or UASCs (unaccompanied asylum seeking children). The UNHCR along with the Separated Children in Europe Programme and the Immigration Law Practitioners Association prefer the term 'separated children'. The Separated Children in Europe Programme prefers to use the term 'separated' because:

> it better defines the essential problem that children face. Namely, that they are without the care and protection of their parents or legal guardian and as a consequence suffer socially and psychologically from this separation. (Save The Children 2004: 2)

This highlights the importance of recognising that separated young people and children may not necessarily be unaccompanied, and may instead by accompanied by a trafficker, sibling or agent. The Immigration Law Practitioners Association (2004: 5) notes that 'separated children' may include, but are not limited to, those who:

- are entirely on their own in the UK;
- are accompanied by a relative who is not their primary carer and who is unable or unwilling to take responsibility for them and abandons them shortly afterwards:
- are accompanied by a person who is or is not a relative who may be caring for them or may be exploiting them;
- are sent by their parents for safety, education, health or other reasons with or without their consent;
- are separated from their families who are in countries other than their country of origin;
- are trafficked against their will.

Along similar lines, the UNHCR (2004) encourages the use of the phrase 'separated' as it is eager to draw attention to the particular protection needs of this group. However, on the other hand, it can be argued that the term 'separated' implies passivity and overlooks the young people's agency, since some of the young people have chosen to move.

with – particular religious, cultural or political organisations or agencies can have profound influences over their experiences and their need to seek asylum.

A key principle of the United Nations Convention on the Rights of the Child focuses upon giving recognition to children's rights to participate in decisions affecting them. As such, Hopkins and Hill (2008) sought to explore the processes behind children's decisions to leave their countries of origin. A range of people participated in this process with young people adopting a somewhat passive role. Many mentioned that their 'uncle' made this decision for them, yet for some their uncle was not necessarily a family member but a friend of the family or someone who lived in the same village. The children's lack of engagement with such important decisions was evidenced by the fact that the majority did not know where they were going, were unaware that they were travelling to Scotland, knew very little about it and were rather disorientated when they arrived. After leaving their country of origin, many children initially visited a neighbouring country before flying to the UK. An agent accompanied the vast majority of the young people and this person was often known to the wider family. Service providers voiced concern about the motivations, behaviours and attitudes of a number of the agents recounting stories of exploitation and abuse. According to this research, there were few accounts of trafficking with some young people – particularly those from the Asian continent – being smuggled into the UK. Upon arriving in Scotland, many separated children recalled being left at the airport, at a central railway station or being sent to the offices of the Scottish Refugee Council. These young people then find themselves in a completely new context where their personal identities mean very different things to how they were interpreted in their country of origin. Furthermore, they also have a new label as an 'asylum seeker' and its associated baggage to carry with them in their everyday lives.

The transition to adulthood

A key paradigm within work about young people, sitting alongside an approach informed by a youth subcultures paradigm, is scholarship that focuses upon exploring young people's experiences of the transition to adulthood (see, for example, Furlong 2009, Furlong and Cartmel 2007, Henderson *et al.* 2007, MacDonald *et al.* 2001).

> The process of moving from total physical dependence to independence is one of the basic underpinnings of how we understand the transition to adulthood. Learning to crawl then walk, crossing the road, walking to school with friends, learning to drive and travelling alone are widely viewed as stages in a story of social development.
>
> (Henderson *et al.* 2007: 101)

Almost all textbooks or edited collections about young people include some discussion about the transition to adulthood, and this suggests that the transition through the lifecourse from being a young person to being an adult is one of the key ways in which researchers view young people. As such, much work focuses on the ways in which young people negotiate these transitions, how these transitions have changed over time and what the key components of such transitions are. Perhaps not surprisingly, this has led to the transition to adulthood being the focus of much debate within youth studies (see for example, Arnett 2006, Bynner 2005, Roberts 2007, Wyn and, Woodman 2006, 2007).

Research about the transition to adulthood tends to focus upon the achieving of particular goals and so often focuses on when young people gain personal independence, an occupation or career, financial autonomy and householder status (Hill and Tisdall 1997). For Jones (2002), the transition to adulthood comprises a number of different components and many young people will have obtained adult status in some of these components and not in others. As such, there is no longer one main transition from youth to adulthood as the pathways are now increasingly diverse. Some young people may need to revisit particular pathways on the journey to adulthood and so the transition is often not linear, and instead comprises intermediate phases. These changes have also altered the significance of particular events insomuch as the move from school to work is often more gradual now than it used to be and so is regarded as a less significant event in the transition to adulthood (see Table 11.1). Although young people are increasingly attempting to become more independent, this process is often accompanied by extending periods of dependency upon others for support, especially given that the transition to adulthood happens in diverse ways:

> For young people in almost all parts of the world, the transition to adulthood is perhaps more complex and contested that in any previous era. The sheer speed of neoliberal economic and social

> reform in many parts of the globe has effected profound changes in young people's experiences. Nations are implicated in a changing global order in which government disinvestment in welfare measures, transnational economic competition, high rates of unemployment, and economic recession is increasing pressures on parents and young people.
>
> (Jeffrey and McDowell 2004: 131)

Drawing upon work within developmental psychology, Arnett (2000, 2001) has advanced the concept of emerging adulthood (Box 11.2) to explain the experiences and circumstances of young people aged 18 to 25. 'For most young people in industrialised countries, the years from the late teens through the twenties are years of profound change and importance' and 'by the end of this period, most people have made life choices that have enduring ramifications' (Arnett 2000: 469). According to young people consulted by Arnett the most important characteristics of the transition to adulthood are 'accepting responsibility for one's self, becoming capable of making independent decisions, and becoming financially independent, in that order' (Arnett 2001: 17). Arnett's approach has been critiqued on a number of grounds (e.g. see Bynner 2005) not least because of its tendency to see emerging adulthood as a phenomenon influencing all young people. Arguably, emerging adulthood is likely to be experienced more by middle-class young people than those who are less affluent, and the role of different economies is likely to shape how the emergence of adulthood is experienced by young people.

Table 11.1 ***Extended transitions to adulthood***

Childhood	*Youth*	*Adulthood*
School	College or training scheme	Labour market
Parental home	Intermediate household, living with peers or alone	Independent home
Child in family	Intermediate statuses, including single parenthood, cohabiting partner	Partner-parent
More secure housing	Transitional housing in youth housing market (e.g. furnished flats and bedsits)	More secure housing
'Pocket money' income	'Component' or partial income	Full adult income
Economic 'dependence'	Economic 'semi-dependence'	Economic 'independence'

Adapted from Jones (2002: 2)

Box 11.2

Emerging adulthood

Jeffrey Jensen Arnett (2000: 469) developed the idea of emerging adulthood:

> I argue that this period, emerging adulthood, is neither adolescence nor young adulthood but is theoretically and empirically distinct from them both. Emerging adulthood is distinguished by relative independence from social roles and normative expectations. Having left the dependency of childhood and adolescence, and having not yet entered the enduring responsibilities that are normative of adulthood, emerging adults often explore a variety of possible life directions in love, work, and worldviews. Emerging adulthood is a time of life when many different directions remain possible, when little about the future has been decided for certain, when the scope of independent exploration of life's possibilities is greater for most people than it will be at any other period of the life course.

For Arnett 'the teens through the mid-twenties are the most *volitional* years of life' (469). Most American youth have left home by the time they reach 18 or 19, and the years that follow this involve diverse residential patterns. 'About one third of emerging adults go off to college after high school and spend the next several years in some combination of independent living and continued reliance on adults, for example, in a college dormitory or a fraternity or sorority house' (p. 471). This is therefore a period of partial autonomy. About 40 per cent of American young people move out of the parental home into an independent living situation and take up full-time employment and about two-thirds cohabit with a partner. This is the highest incidence of residential change compared with all other social groups. According to Arnett, important criteria for having reached adulthood include taking responsibility for yourself, making autonomous decisions and being financially independent. Emerging adulthood is a period of identity exploration in terms of love, work and perspectives on the world.

Adapted from Arnett (2000)

Gill Jones's (1995) work about young people's experiences of leaving home (explored in Chapter 5) is also very relevant to discussions about the transition to adulthood. 'Leaving the parental home is a normal part of the overall process of transition from dependent childhood to independent citizenship' (Jones 1995: 1):

> Young people's emancipation from parental control – an essential part of the transition to adulthood in western industrialised societies – therefore involved moving out of the home of their parents and establishing a home of their own, for which they have responsibility and in which they have power and control.
>
> (Jones 1995: 3)

This transition is often experienced in diverse ways and may be accompanied by insecurity, unhappiness and abuse, or it may be associated with excitement and independence. By leaving home, young people are 'affectively acknowledging the responsibilities and claiming the rights of adult citizens' (Jones 1995: 4), yet many young people leave home with insecure futures and limited job opportunities.

Just as the transition to adulthood connects with home, so too it is often interrelated with other scales and senses of belonging and attachment. 'Our passages, intensely personal, thread their way through, impact upon and are influenced by the institutional fabric of social life: home, work, school, family, religion, nation' (Kenworthy Teather 1999:1). Another example of the ways in which young people engage with key transitions in their lives can be found in Winchester *et al.'s* (1999) study of 'Schoolies Week' on the Australian Gold Coast. This week is often regarded as a rite of passage from youth to adulthood, where school leavers spend a week away from home and school. The geography of the week is significant as 'it occurs in a highly constrained period of space and time, and involves ritualised and transgressive bodily experiences. The spatial context is significant' (Winchester *et al.* 1999: 59). The actual physical separation from their homes and places of the everyday allows these young people to detach themselves and transform their identities into something new. The freedom of choice allows the young people to reject any form of structure and thereby show that they are no longer tied into the limits of school and youth. The spatial context means that the young people can do what they want and when they want without the 'accountability of parental or school supervision' (Winchester *et al.* 1999: 67).

The gap year

The focus of this final section about migration, mobilities and transitions is the increasingly common phenomenon of taking a year out, often referred to as a gap year. This is generally regarded as a one-off opportunity to spend time travelling and exploring different places. 'Once a marginal and unusual activity undertaken by hippies and adventurous drop-outs, it has become a widely accepted rite of passage for young people' (O'Reilly 2006: 998) although it is still likely to be accessible to only an elite group of young people. As Camille O'Reilly observes, having a gap year is now recognised as a key component of young people's transitions to adulthood as they negotiate the processes of becoming independent whilst also having fun and discovering different parts of the world. Estimates of the number of young people participating in a gap year include over 200,000 young people aged 18–25 (Simpson 2005), with 350,000 overseas visits being offered annually in 200 countries. Furthermore, 6 per cent of university entrants have previously taken a gap year, most in the Third World and lasting between three and twelve months (Ansell 2008). As O'Reilly (2006) explains, people's engagement with backpacking or taking time out varies depending on the length of time they travel for, the form of transport they use and the extent to which their trip is planned or not. Some may participate in a pre-arranged programme through voluntary groups, travel companies or religious organisations and may not be involved in any pre-planning or organisational activities whilst others drift: 'the drifter ideal is one of little or no advance planning, allowing word of mouth and serendipity to influence the itinerary' (O'Reilly 2006: 1001).

As Kate Simpson (2004, 2005) has shown, the gap year has become increasingly professionalised in recent years as youth travel practices have adopted an increasingly corporate focus. The result is that:

> An industry has emerged around gap year practices which functions to formalise, popularise and to some degree police the industry's activities. Fundamental to these processes has been the emergence of a set of marketable values that can be accepted by institutions, desired by parents, and bought by young people.
>
> (Simpson 2005: 450)

An important commodity gained through participating in a gap year relates to the cultural capital obtained by young people alongside the access to employment and other social opportunities that this often leads to (Heath 2007). Companies tend to brand gap-year activities as socially enhancing, connected with success and fundamental to a good CV. As such, taking part in a gap-year is being marketed as a resource that young people require in order to successfully negotiate the transition to full employment and adulthood.

In terms of young people's motivations for travel, Camille O'Reilly (2006) observes that the changing nature of the workplace and, in particular, the increasingly insecure nature of work, means that it is now easier for people to decide to delay starting work, or to stop working in order to travel without being overly concerned about gaining employment in the future. In her research, O'Reilly (2006) noted that those working in information technology were very confident about leaving work behind as they felt it would not be too difficult to find work when they returned. As well as the changing nature of work, other factors that have influenced the increasing engagement with travel include cheap international air travel and the tendency to value such experiences rather than seeing them as being exotic or decadent.

Linking travel with identity, Luke Desforges (2000) has discussed how many of the travellers he interviewed (which included some young people in their twenties and early thirties as well as older participants) mentioned that their decision to travel was often closely connected to points in their lives when they were questioning their self-identity. As such, it could be argued that young people's engagement with backpacking, travelling or taking a gap year is partly a response to their questioning of their personal identities and future trajectories. Furthermore, for some of the young participants, gaining 'new experiences of the world is closely tied to a youthful identity' (Desforges 2000: 936). For them, understanding what it means to be young is closely connected with these experiences and journeys.

In another example, and drawing attention to issues of risk, Nicola Ansell (2008) has discussed the ways in which gap-year projects which take place in the Third World place young people in distant environments which are often perceived of as being risky. She analysed texts and images on the websites of gap-year providers as well as interviewing representatives of two organisations and conducting focus groups with young people who had participated in a gap-year. Ansell (2008: 227) observes how gap-year providers must simultaneously highlight the 'challenge and adventure' of participating in a gap year, whilst balancing this out with selling the gap year experience as one that is safe and secure. This is particularly crucial because in terms of intergenerational relations, it is estimated that parents are the main source of financing a gap year for 21 per cent of young participants with 17 per cent of parents attempting to persuade their children not to go on a gap year (Ansell 2008). In exploring this issue, Ansell (2008) detects multiple strategies for managing risk including: the marketing of risk as adventure; the marketing of relative safety; the use of risk assessments; providing volunteers with advice or treating them as 'not quite adult' and involving parents. As she concludes:

> Gap year projects . . . commodify a form of youth transition that is valuable to middle-class youth and that permits intergenerational transmission of class in an individualising world. Gap years mediate between modernising, risk-averse trends in society (expressed particularly through parents) and postmodern, risk-embracing, individualising trends (expressed through youth), not by adopting a midway compromise but by selectively representing and controlling elements of spatially uneven risk. This enables young people to draw on the resources of both parents and of Third World environments to forge distinctive identities in late-modern Britain.
>
> (Ansell 2008: 237)

Overall then, it is clear that gap years interconnect with young people, place and identity in important ways. Contributing towards the transition to adulthood, shaped by intergenerational relations and engaging with different places, participating in a gap year is important to young people's aged identities as well as influencing their identities as potential employees with particular resources and personal qualities.

Key themes

- Although migration and mobility are key characteristics of many young people's lives, there is a lack of research about the ways in which young people's everyday lives shape and are shaped by migration and mobility.
- Many young migrants are forced to move as a result of personal or family persecution resulting in them becoming asylum seekers in other countries. This often results in stark changes to their senses of identities, the ways in which they receive particular identity labels along with their negotiation of social identities more generally.
- The transition to adulthood is a key mechanism through which young people are understood and this transition has diversified and became extended in recent years.
- Participating in international travel through taking a gap year has become increasingly professionalised in recent years resulting in it being promoted as a necessary commodity for young people to successfully negotiate the transition to adulthood.

Project ideas

Critically discuss the complexities of what it means to have the identity of a young asylum seeker or refugee. What are the key components of such an identity, who defines this and what processes are in operation to arrive at such definitions?

What are the main components of the transition to adulthood? Are some more important than others? Why?

Would you prefer to participate in a pre-arranged gap year or one that is more flexible? Why?

Suggested further reading

This article is a very useful account of the ways in which researchers explore the intersections of youth and mobility:

Barker, John, Kraftl, Peter, Horton, John and Tucker, Faith (2009) The road less travelled – new directions in children's and young people's mobility. *Mobilities* 4(1) 1–10.

This article explores issues associated with gap years and the ways in which it creates inequality amongst young people:

Heath, Sue (2007) Widening the gap: pre-university gap years and the 'economy of experience'. *British Journal of Sociology of Education* 28(1) 89–103.

This is a brilliant book that explores the experiences and accounts of young people's transitions to adulthood from a biographical perspective:

Henderson, Sheila, Holland, Janet, McGrellis, Sheens, Sharpe, Sue and Thomson, Rachel (2007) *Inventing adulthoods: a biographical approach to youth transitions*. London: Sage.

This useful article discusses the pre-flight experiences of unaccompanied children as well as their accounts of migration and arrival:

Hopkins, Peter and Hill, Malcolm (2008) Pre-flight experiences and migration stories: the accounts of unaccompanied asylum-seeking children. *Children's Geographies* 6(3) 257–68.

This excellent article argues why youth transitions research is of value to youth studies:

MacDonald, Robert, Mason, Paul, Shildrick, Tracy, Webster, Colin, Johnston, Les and Ridley, Louise (2001) Snakes and ladders: in defence of studies of youth transition. *Sociological Research Online* 5(4) www.socresonline.org.uk/5/4/macdonald.html.

This article explores contemporary issues associated with the gap year, in particular the professionalisation associated with it:

Simpson, Kate (2005) Dropping out or signing up? The professionalisation of youth travel. *Antipode* 37(3) 447–69.

12 Urban–rural

John: I get lost downtown. There's too much going on down there. Too many lanes. You go in there and there's people and you're flying and somebody jumps in front of you. No, I get mad so I just don't go down there. (Vanderbeck and Dunkley 2003: 252)

Christopher: If we've got the money, we'd go and play pool and have a few pints. (McDowell 2002: 112)

Francis: I don't want to work for a company that I'm not – I don't think I'll be happy in. So, I'm willing to postpone things and make sure I get the right one. (Henderson *et al.* 2007: 52)

Young urban–rural places

The quotes above from John, Christopher and Francis offer some explanations for their experiences of urban and/or rural places. John's preference is to live in a rural area as he finds the city a busy and bustling place with lots going on. Contrary to this, Christopher's reflection on his life in the city is about what he does in his leisure time: play pool and have a few drinks. Francis talks about looking for work in the city and prefers to wait for a job he is happy in rather than taking up work that won't satisfy him. In different ways, the perspectives of these three young people draw attention to specific aspects of growing up in urban and/or rural places. There is a general tendency within public and academic discourse to see urban youth and rural young people as completely separate social entities. The sense is that experiences of growing up, accounts of everyday experiences, and processes of identity formation are very different for urban young people compared with their counterparts living in rural areas. These understandings are regularly reinforced by very powerful stereotypes about rural and urban areas. Picturesque landscapes, the natural environment of the countryside, open spaces and fields all contribute to the stereotype that rural places are ideal locations for children and young people to grow up. Rural communities are also often stereotyped as harmonious, close-knit places away from the stresses and strains of urban life (Kraack and Kenway 2002), and as such, the rural is therefore often read in opposition to the urban which is represented as stressful, threatening and dangerous. As Scott, Gilbert and Gelan (2007) highlight, stereotypes about urban and rural populations can be mapped across a range of dimensions of social relations including education and employment, ethnicity, migration and sense of community (Table 12.1).

Clearly these stereotypes of urban and rural life are problematic for many reasons, particularly in the ways in which they homogenise all that is rural and all that is urban. Just as there are many experiences of urban living, so too, there are many different ways in which young people may be classified as rural or identify with rurality. Urban young people encompass those living in busy inner cities or in leafy suburbs, housed in an apartment, a large detached house or a high-rise block. Similarly, rural young people may live in a market town, a small village,

Table 12.1 ***Stereotypical differences between urban and rural populations***

Dimension	*Urban*	*Rural*
Economy	Secondary and tertiary sector dominant	Primary industry sector and supporting activities dominant
Employment	Manufacturing, construction, administration and services	Agriculture, forestry and other primary industry occupations
Education	Higher than national averages	Lower than national averages
Service accessibility	High	Low
Information accessibility	High	Low
Sense of community	Low	High
Demography	Low fertility and mortality	High fertility and mortality
Political views	Liberal and radical element more strongly represented	Conservative, resistant to change
Ethnicity	Varied	White
Migration	High; generally net in-migration	Low: generally net out-migration

Adapted from Scott *et al.* (2007: 4)

a hamlet or a very remote farmhouse. Added to this, the identities of these young people are read, negotiated, conveyed, employed and performed in different ways resulting in a complex array of different experiences and accounts of being an urban or rural young person.

Although much academic work seeks to trouble what is meant by urban or rural, there are strong traditions within contemporary social science research that can broadly be identified as urban studies and rural studies. These traditions are produced and sustained through the publication of books and edited collections focusing specifically on the urban or the rural, separate journals for Urban Studies and Rural Studies and distinct research groups, conferences and workshops. As such, there is a sense in which the production of academic knowledge works to create and sustain the urban and rural as separate despite the linkages and connections between them and the blurry boundaries between what might be constituted as urban or as rural.

Having said this, it cannot simply be concluded that urban and rural are not useful categories of analysis. For example, clearly the young people researched by Cahill (2004) in New York are growing up in very different contexts from those with whom Dunkley (2004) conducted research in northern rural Vermont. Whether it is the density or diversity of population, the sense of open space or the availability of services, there are stark contextual differences between these two research locations. However, despite the separate treatment of the urban and rural, these two studies also highlight many of the commonalities experienced by young people in urban and rural locations. Whether is it the negotiation of adult demands, the search for appropriate places to hang out or the management of exclusionary processes associated with identity politics, young people in both contexts also have much in common with each other. Panelli (2002: 119) has identified a range of strategies for researching the lives of young people negotiating rural life, posing a number of strategic and searching questions for academic research (Table 12.2). Although clearly applied to the experiences of rural young people, these negotiations and questions could all be applied to urban young people as well, highlighting the connections and interrelations between what is constituted as rural or urban.

In this chapter, I consider the urban and rural simultaneously in order to highlight both the connections and associations as well as the differences and disjunctures between the urban and rural in the context of young people, place and identity. I do so for two main reasons. First,

Table 12.2 ***Negotiating rural life: strategies for investigating young people's lives and experiences as dynamic***

Negotiations	*Strategic questions*
Negotiating rural knowledge	What rural knowledge do young people hold? How and where do they gain cultural understanding of rurality? How do young people produce their own meanings and knowledge of rural life?
Negotiating rural work	What productive and reproductive work do young people do in rural settings? How do young people learn and negotiate tasks? How does work shape young people's rural experiences, both socially and spatially?
Negotiating rural social relations	Where are young people located in rural units: families, households, communities? What social relations and differences shape young people's lives (including their movement) in rural settings? How do young people engage with, contest or change the social relations in which they participate?
Negotiating political participation in rural settings	What are young people's experiences of decision making and political participation in rural contexts? How can young people respond to established rural policies and political structures? How can institutions and political processes be changed to recognise more appropriately and include rural young people?
Negotiating rural space and place	How do young people experience and understand different rural spaces and places? What are the relations between space, place and movement for young people in different social, economic and cultural contexts? How do young people mobilise particular rural spaces or construct places of significance for their own lives?

Adapted from Panelli (2002: 119)

having two separate chapters on the urban and the rural reinforces simplistic and problematic stereotypes and assumptions about each, particularly with relevance to young people, place and identity. By considering the urban and rural together, I hope to demonstrate the relations between them as well as the ways in which these contexts also offer unique experiences for young people. Second, although clearly the landscapes experienced by urban and rural youth differ in quite marked ways, research has shown that the separation of rural youth from urban youth is problematic given that many rural youth are connected to a variety of communities – from the local to the global – in a range of different ways (White and Wyn 2008). The content of the previous chapters connects closely with this chapter as urban and rural youth people all perform, negotiate, convey and employ particular identities in ways which connect with their embodiment, their experiences of home, their negotiations of public space and their feelings about the neighbourhood and community.

One of the few examples of scholarship which seeks to demonstrate the connections between the urban and rural for young people is Vanderbeck and Dunkley's (2003) research about young people's narratives of urban–rural difference in the USA. They argue that there has been a lack of attention to the ways in which understandings of urban–rural differences act as sources of identity for young people. Drawing upon two case studies (one urban and one rural) from the USA, they argue that young people often draw upon broader public discourses and invoke hierarchies of places, people and activities which often represent the urban as better than the rural. This demonstrates the importance of remembering that the process of identification is often about how people identify as well as how they dis-identify.

Vanderbeck and Dunkley (2003) observe the powerful sets of stereotypes and everyday discourses that exist which associate children with rurality and the rural idyll, with the rural being represented as the ideal place to bring up children. The tendency is to believe that 'children are innately closer to nature and will flourish in a rural environment where they can breathe fresh air, share proximity to animals, and have freedom to roam' (Vanderbeck and Dunkley 2003: 246). This set of assumptions may be reinforced by parenting practices but is also mentioned regularly in children's stories, films and television programmes. They critique the discourse of the rural idyll not only for its stereotypical and idealistic view of the rural but also because it has a questionable applicability to understandings of rurality in many places in the USA. For example, they observe that African American youth growing up in the rural south are unlikely to see rural life as complying with idealistic notions of rurality, given their experiences of racism, poverty and the stigma associated with their southern accents. They also draw attention to the importance of recognising the diversity of places that are classified as urban or rural, observing that particular regions, territories or even nations are often problematically labelled as 'rural' or 'urban' in stereotypical ways.

The young people consulted by Vanderbeck and Dunkley (2003) drew upon a number of reference points in constructing their narratives of rural–urban difference. This included young people's own personal experiences, migration histories as well as their interpretations of everyday discourses on television, through social networks or through schooling. Many of the participants were somewhat ambivalent about the extent to which they identified as urban or rural, leading the researchers to suggest that there is a need to rethink the labels of urban and rural: 'there is considerable scope . . . to take more seriously the urban–rural continuum alongside other facets of identity that have received critical attention, as well as to explore the intersections of these facets of identity' (Vanderbeck and Dunkley 2003: 256). In the context of these debates, I simultaneously consider in this chapter the experiences of young people growing up in rural and urban areas focusing specifically upon places of leisure, drinking and work. In each of these three sections, I highlight the diversities within young people's urban and rural experiences as well as the commonalities and differences articulated by young people through their accounts of growing up in a variety of urban and rural contexts. As such, an important theme of this chapter is also the relationalities between the categories of youth, adult, urban and rural.

Leisure places

An important theme within research about young people's urban and rural experiences focuses on the places they use and engage with during their leisure time. There is a range of leisure places that young people use across the urban–rural continuum from petrol stations, internet cafés and street corners through to shopping malls, playing fields and youth clubs. What is particularly interesting is the way in which conflicts over the use of leisure space are issues in the majority of rural and urban contexts as young people compete with other members of society in attempts to claim space. Furthermore, it is also evident from research that young people respond to these conflicts in a range of diverse ways as well as contesting dominant understandings of what it means to be an urban or rural young person.

Tucker (2003) conducted research with 27 girls aged between 10 and 15 about their experiences of growing up in rural Northamptonshire in the UK. She observed that many settlements in rural areas often have limited provision of leisure facilities, such as playgrounds, youth clubs and other such sites associated with young people. Alongside this, informal spaces in the countryside have often been privatised, restricting the places available for young people to participate in leisure activities with their peers. These restrictions result in young people having to share recreational spaces with adults as well as with other groups of young people, and also often results in increasing the visibility of young people making it more likely that they will be subject to surveillance by adults. Tucker (2003) found that many of the girls she worked with had experienced conflict with adults and that this was often heightened as adults viewed the young women as a problem when they were with their friends. Girls recollected being reprimanded by adults for making too much noise, questioned about what they were up to and generally made to feel unwelcome. These experiences led girls to respond either through compliance, retreat or resistance. It is likely that many young people living in urban contexts will experience similar forms of conflict and surveillance; however, Tucker (2003) observed that many of the girls she worked with were known by name by local people. This is perhaps an aspect of some young people's experiences of rural living in that they may be more likely to be known locally rather than being afforded the anonymity that some urban young people may feel they have.

Drawing upon the concept of intergenerationality, Tucker (2003: 117–18) states that 'what emerges is that generational differences – that is, differences in attitudes and actions between groups who draw upon shared experiences with other people of a similar age – may result in disparate understandings of appropriate place use and behaviour'. However, she also notes that the girls' responses to these issues are diverse and they employ a range of tactics to manage the demands placed upon them by adults (Tucker 2003). Furthermore, although young people may conflict with adults in rural areas, they may also conflict with other groups of young people, especially so when there is restricted space available. Tucker (2003) found that dominant groups sometimes actively sought to exclude young people through intimidation and bulling, whilst some young people chose to avoid certain places so as not be associated with or come into contact with particular groups of young people. Some young people also felt that the presence of adults often minimised the likelihood of there being any conflict. Another area for conflict, identified by Tucker and Matthews (2001), may arise not only between different groups of young people but between girls and boys. They found that, as in many urban locations, girls were often somewhat marginalised due to the presence of boys and were frequently governed and regulated by boys in terms of how they used different places. Playing fields and recreation grounds were even labelled by some girls as 'boy places' (Tucker and Matthews 2001: 166). Clearly then, 'growing up is a varied experience comprising multiple realities of difference and diversity' (Tucker and

Matthews 2001: 162), involving different contestations over the use of space and different types of response from young people.

Drawing upon her work in Norway, Laegren (2002) observes that petrol stations are often the only places in rural Norway where young people can meet their peers. These sites are often places for young men to engage in car culture activities, as well as heavy drinking. Laegren (2002) describes petrol stations as the territory of 'råners' – young boys, aged 15–22 who are interested in cars and who are often accompanied by their girlfriends. Although the petrol station is an important place, Laegren (2002) notes that the car is also an important commodity for these young men as they convey, perform and construct their identities. A 'rane' car is easily recognised by the young men – 'it is slightly lowered and has new wheel-trims, dark-shaded windows, white spotlights, spoilers and various bits of extra equipment' (Laegren 2002: 161). The majority of the young men do all of the work on the cars themselves, building on skills they have developed through their friendship networks. The practices of these young men – hanging around at the petrol station and driving their cars (sometimes at very high speeds) – is seen as a problem among many adults and other young people, and also challenges simplistic understandings of the rural. It could be argued that their behaviour is transgressional and this work has strong connections with the boy racers discussed in Chapter 10.

As an alternative to the petrol station, the growth in internet cafés over the last five to ten years has provided additional spaces for young people to hang out. The research conducted by Laegren (2002) focused upon an internet café in a rural village in Norway that opened after local young people and adults led a voluntary initiative supported by local companies and organisations. For some, the internet café offered the young people a taste of city life. The café was open in the early evening and was staffed by one adult and two young people. Young people browsed the internet, adults had coffee, and on Wednesdays older people would meet to play bingo, so the internet café became an important location of intergenerational interaction. It was an important place for the young people, not only to access the internet but to meet up. Also, the users of the internet café often had contacts outside the village and were seen as more cosmopolitan than other young people. Most of the regulars at the café identified with 'frik', a particular style of youth culture – seen as in opposition to mainstream fashion and connected with the alternative arts and music – often associated with the urban areas of Norway. In this example, it appears that a particular group and type of young person is dominating the use of the internet café, preventing other groups of young people from using this facility.

Leyshon (2008b: 4) contends that the different inclusions and exclusions experienced by rural youth are poorly understood:

> [I]t is important to begin by conceptualising young people as active agents in their own decision making and identity formation and to view their spaces in terms of an open complexity of interacting social relations . . . it is also important to recognise . . . that, within that open complexity, adults, young people and social groups are constantly engaged in efforts to territorialise, to claim spaces, to include some and exclude others from particular areas.

Leyshon (2008b) conducted research with young people – in the rain, in pubs, at concerts or in the corners of fields – working in collaboration with detached youth workers in rural areas. The young people determined the methods to be used and so some employed photography, some video diaries, some group discussions and others group film-making. All the young people identified as rural youth and were aged 16–19. Leyshon (2008b) found that the rural young people he worked with revered the spaces of the local village as key places in the ways in which they constructed rural identities. Alongside this, the fields, woods, hedgerows, farms and lanes were also important aspects of rural life according to the young people as they are the context for the how the rural works, in terms of agriculture, the environment and production more generally. Several young people felt that the rural was peaceful, beautiful, healthy, thereby complying with dominant discourses about rurality. However, the young people also tended to identify personally with these qualities too, making a strong connection between the qualities of the countryside and their own personal identities.

The young people both celebrated and derided the ways in which rural villages are constructed as close-knit places where everyone knows each other. Some of them found these qualities of the rural repressive and complained about the muddy fields, the open spaces and nosy local people. Some also found that most places came under adult surveillance, restricting the opportunities they had to engage with the opposite sex, and others mentioned the limited pool of people available to them for potential relationships in their local area. This can be particularly challenging for young people who do not conform to heteronormative assumptions about appropriate everyday behaviours, which are felt particularly strongly in rural areas. Leyshon (2008a) discusses the importance of considering how young people's experiences of the countryside are shaped by memories of place and experiences of rural society, an important part of which revolves around and is reinforced by stereotypes of the rural idyll. He suggests that there is a moral geography of the countryside that young people negotiate which often involved the exclusion or explicit engagement with the urban world and its associations with modernity. Indeed, being a rural youth is not just about being in a rural location but involves the adoption of a moral code associated with appropriate rurality.

The rural young people participating in Leyshon's (2008a, b) research were also imaginative and creative in finding and providing entertainment opportunities given the relative lack of amenities and opportunities, and there was a general sense amongst the young people that the rural is under threat from the urban. So, although rural youth engage in cultural activities that might be associated with the urban – through books, magazines, the internet, TV – they also construct the urban as different, inferior and unsophisticated compared with the rural:

> For rural youth there is a betweenness of being in the countryside. Whilst they enjoy 'urban' entertainments they are keen to mark themselves out as different to urban youth and variously locate themselves both within the countryside and beyond.
>
> (Leyshon 2008b: 22)

These studies about the experiences of rural youth clearly demonstrate the diverse ways in which young people experience rural areas. Key issues include conflicts over other social groups – adults, other groups of young people, and conflicts between young men and young women – about the use of different leisure spaces. Moreover, different sites within rural areas can become associated with particular groups of young people, resulting in other young people experiencing marginalisation and exclusion, either subtly or blatantly. Furthermore, young people simultaneously reinforce and challenge dominant understandings of the rural through their use of place and the construction of their identities, often drawing upon references associated with urban experiences.

A number of the examples drawn upon in previous chapters about neighbourhood and community and public space and the street draw upon research with young people in urban contexts. It is interesting to observe that although the rural is often constructed by young people in relation to the urban, there appears to be less evidence that urban young people use the countryside as a reference point for how to construct their experiences of cities. Just as Leyshon (2008a, b) observes that young people refer to the fabric of the countryside – lanes, hedgerows, fields – as important to young people's senses of identities, so it may also be the case that some urban youth value the streets, buildings and urban fabric of their lives as key components of how they construct their identities as urban.

For many young people living in inner cities or suburban neighbourhoods, shopping malls offer them space to meet up with friends (e.g. O'Dougherty 2006). Vanderbeck and Johnson (2000) conducted research with 12- and 13-year-old African American young people in an economically disadvantaged inner-city area in the south-east of the USA. All of the young people raised concerns about violence and personal safety with regard to their local neighbourhoods; they tended to find their local area boring and were frustrated by the lack of opportunities in terms of places available to play with friends as well as by personal and parental concerns about heightened risks in public spaces. For many of the young people, the shopping mall offered them an appealing alternative to the restrictive places of their home and local community. They regarded the shopping mall as offering increased mobility as well as the opportunity to make individual choices about where to go and with whom. Young people were attracted to the consumption opportunities of the mall as well as the stimulation it offered them.

A key concern in a range of studies of urban and rural youth is the lack of space available for young people to develop their own personal identities. Giddings and Yarwood (2005) found this in their research with young people in rural Hampshire, UK, and many of the studies mentioned previously highlight this as a concern as well. Moreover, there is also evidence of adults regulating the use of the public spaces chosen by young people to socialise with friends, forcing them into other spaces, often further from home (Giddings and Yarwood 2005). It could be argued that these restrictions placed on young people's use of space by adults result in young people having less space to experiment with their identities and senses of self (Aitken 2001a, Tucker 2003). This relates to the idea that young people are increasingly experiencing what Katz (1998) calls 'eroding of ecologies of youth' whereby young people's spaces are suffering due to increasing levels of disintegration and disinvestment resulting in them having a lack of appropriate facilities and having very little space to perform their identities:

> Young people's growth and development depends upon environments that provide stimulation, allow autonomy, offer possibilities for exploration, and promote independent learning and peer group socialising. These criteria are important in all settings, not just those designed specifically for teens such as schools, leisure environments, and teen centres.
>
> (Katz 1998: 141)

Drinking places

> It is a cold Friday night, and windy. You are dressed for dancing, not the weather. Finally, you reach the head of the queue outside the night club. The bouncer – although nowadays they prefer to be called doormen – raises his arm and lets your friend in. He takes one look at you and demands proof of age. All you have in your pocket is money. That is not enough.
>
> (Jenkins 2004: 1)

A regular and important activity for many young people involves participating in nightlife spaces and drinking places, a set of activities which are overwhelmingly concentrated in urban areas. As Jenkins (2004) highlights in this statement, engaging in nightlife spaces often involves a range of stages and activities from meeting with friends, joining the queue, gaining entry and so on. Although this may be the perception of many young people on a night out, Chatterton and Hollands (2002) propose that urban nightlife can be theorised through consideration of the ways in which it is produced, regulated and consumed (Table 12.3). They observe that although many cities in the UK started to experience urban decline from the 1970s onwards with unemployment, social decay, crime and homelessness being widespread, within the last two decades many have been centres of economic development, focusing particularly upon leisure, retail and consumption-based forms of industry. This focus upon urban cultural economies has been accompanied by the growth and development of nightlife economies in many cities, mostly centred

Table 12.3 ***Urban playscapes***

	Type of nightlife space		
Mode of analysis	*Mainstream*	*Residual*	*Alternative*
Production	Corporate brand Profit-oriented Global/national owned companies	Community Need-oriented National/regional	Individual Experimental Local
Regulation	Entrepreneurial Formally controlled (CCTV, bouncers etc.)	Stigmatised Formal (police)	Cautious Informal (self-regulated)
Consumption	Profit-oriented Divided consumer-producer relations (brand/lifestyle) Sanitised/up-market	Community-oriented and focused Traditional consumer-producer relations (product) Down-market	Creative-oriented Interactive consumer-producer relations Alternative/resistant
Location	Dominant centre	Underdeveloped centre	Marginal places

Adapted from Chatterton and Hollands (2002: 100)

on bars, clubs, pubs and music venues. In order to gain a full appreciation of the complexity of nightlife spaces and drinking places, I now consider the production, regulation and consumption of nightlife spaces.

In terms of the production of nightlife spaces in the city, Chatterton and Hollands (2002) found that the ownership of outlets tends to be controlled by larger national and multinational companies, with as many as two-thirds of nightlife spaces in large cities being controlled by national operators. Many nightlife spaces were renovated during the 1980s and 1990s with the aim of meeting the changing demands of assumed variations in the tastes of consumers. These changes, accompanied by regulatory adjustments, have resulted in increasing levels of branding as nightlife spaces are produced for particular markets such as students, young professionals or sports groups. In terms of the implications of the changes for young people, Chatterton and Hollands (2002) observe that choice has increased, especially given the decline in male-dominated drinking places. However, these changes in the production of nightlife spaces also mean that such environments offer a 'largely standardised, sanitised and non-local consumption experience' (Chatterton and Hollands 2002: 102). Furthermore, these changes have had a questionable influence over binge-drinking cultures and the maximisation of opportunities for women to visit such places without being harassed (Chatterton and Hollands 2002). Moreover, these places tend to target young professionals, graduates and students, with local groups or less affluent young people being marginalised or excluded from participating due to price, dress and style. Overall, the changes limit the nightlife opportunities available to young people, especially if they have difficulty in accessing alternative spaces located in the urban margins.

There are a number of different dimensions to the ways in which nightlife spaces are regulated, including 'legal (laws and legislation), technical (CCTV and radio-nets), economic (drinks and door entry prices) and social-cultural (music taste, youth cultural styles and dress codes)' (Chatterton and Hollands 2003: 47). Furthermore, there has also been a shift in regulatory processes from the police and magistrates towards local authorities and their interests in promoting the cultural growth of cities and regions:

> City centres, then, are increasingly being deregulated with respect to the cultural economy and aiding corporate investment in the licensed sector, while at the same time young adults are experiencing greater social and spatial control through formal mechanisms such as CCTV and informally through pricing and dress codes.
>
> (Chatterton and Hollands 2003: 105–6)

Ultimately, then, the consumption practices of young people interrelate with the ways in which nightlife spaces are produced and regulated. Indeed, it could be argued that drinking and engaging in the nightlife economy are a key part of how many young people identify themselves and are identified by others. Jo Lindsay (2009), drawing on research with young Australians, has found that self-control is a major theme that young people discuss with reference to alcohol consumption but argues that a desire for self-control is challenged by cultural expectations associated with heaving drinking as well as exploitation by the market.

The dominant form of consumption for young people often involves mainstream nightlife locations which are characterised by 'smart attire, commercial chart music, circuit drinking, pleasure-seeking and hedonistic behaviour' (Chatterton and Hollands 2002: 109), often within corporately owned bars or clubs. Some mainstream venues target the student market through student nights and drink deals, whilst others aim for a more affluent group through upgrading their interiors and offering designer drinks. These places offer a more exclusive nightlife experience where young professionals and affluent students can access experiences separate from those patronising the cheaper mainstream venues. Moreover, young people's experiences of engaging in drinking places may also be a focus of their peer group relationships and sense of how they create and shape their identities. Sebastian Tutenges and Morten Hulvej Rod (2009) have reflected upon the popularity of young people's drinking stories. Drawing upon research with Danish youth, they note that young people's drinking stories may be vulgar and include accounts of vomiting and urinating. However, they also argue that these stories provide an important platform from which young people can construct their identities, entertain their peers and explore everyday behaviours.

Drawing upon research in rural northern England, Valentine *et al.* (2008) explored cultures of alcohol consumption amongst young people in rural areas. Although binge drinking has been an increasing concern in the UK for some years now, it has always been constructed as an urban problem, heightened by the ways in which some cities market themselves as centres of consumption. Despite the overwhelmingly urban focus of concerns about binge drinking, this research found that many staff working in rural agencies and services found similar issues in rural locations, and there were accounts of some young people, aged 12, 13 or 14, being given alcohol by their parents.

The young people consulted by Valentine *et al.* (2008) all recounted participating in underage drinking when they were younger (as indeed did many of the older research participants). Some of this took place at home with their parents, such as on a special occasion or during a family meal. All of the young people consulted described participating in binge drinking after turning 18, although they did not show any particular concerns about health and drinking. The young people also recollected their parents permitting them to drink with their friends before or instead of going out. They understood the flexibility of their parents as being motivated by perceptions that the home is a safe place to consume alcohol and also recalled being allowed to consume alcohol whilst under age in other situations, such as in rural leisure spaces (e.g. local fairs, cricket clubs and young farmers' events). Although some drinking took place with the knowledge of parents, the young people also described using public spaces or commercial locations to consume alcohol. Some would use bus shelters or other similar places; some would negotiate with older siblings or friends to buy alcohol for them in pubs or other venues; and some would buy only soft drinks but mix this with pre-purchased alcohol having gained entry to a particular venue.

Although both young men and young women were interested in controlling their drinking, women were concerned most about appearing to be sexually promiscuous whilst men were anxious about being tormented for not being able to manage their drink. Furthermore, although women now have far more access to drinking in public than in previous generations, it was clear from the research participants that drunken behaviour by women was viewed negatively compared with men's drunkenness. Furthermore, the young women interviewed argued that young women were more sensible than young men in knowing when to stop drinking.

Although many young people participate in drinking cultures regardless of their urban or rural location, there are some distinctive aspects of the nightlife experience depending on context. Unlike in cities where there are often themed bars and clubs, Valentine *et al.* (2008) found that young people in rural northern England had to share spaces with a mixture of different social groups and ages as there were not enough places to have specifically youth-focused drinking venues. This could work to control young people's behaviours due to the risk of friends, acquaintances, family friends or others witnessing their activities. Furthermore, another of the crucial differences between rural and urban young people's experiences of drinking cultures comes towards the end of the night. The majority of urban youth have access to night buses or taxis whilst rural youth living in outlying villages have restricted access to transport and may have to book taxis in advance. Some rural young people may also take turns driving, thereby limiting their intake on particular evenings (Valentine *et al.* 2008); this is less likely to be necessary for urban young people.

Young people in urban and rural contexts may also use informal spaces, such as drinking at home with their friends before going out, enabling them to be sociable whilst managing costs. However, in rural areas, after closing time young people often also use parks, cemeteries, riverbanks and bus shelters to consume alcohol, sometimes purchased from a nearby pub before it closed for the evening. Furthermore, although the young people consulted by Valentine *et al.* (2008) participated in late-night public drinking, there was little evidence of them behaving in a disruptive manner and they were likely to be small groups rather than in crowds. According to Valentine *et al.* (2008) this is possibly one of the reasons why binge drinking in urban areas appears less of a problem in rural areas. There are two aspects relevant to young people, place and identity that are pertinent here. First, young people in rural areas tend to have access to public spaces that are relatively isolated and sheltered from local adults, meaning that they are often able to find places where they can consume alcohol without interfering with or disturbing others. Second, unlike in many urban locations where young people are reasonably anonymous, rural young people are more likely to be known by others living in their village or town and this lack of anonymity or possibility of being identified may encourage them to monitor and control their own conduct.

Dunkley (2004) explored the experiences of Vermont teens, especially young men, who travelled to Quebec bars in order to escape their childhood and participate in adult activities such as drinking. Dunkley (2004) was motivated to explore this issue due to the fact that many young men lose their lives driving (often returning from Quebec) and participating in activities such as high-speed driving, drug use and drinking. Vermont had the highest rate of young people killed due to alcohol-related motor vehicle accidents during the research and had held this position for the previous three years. Using interviews, discussions and mapping exercises, supported by questionnaires conducted at local high schools and participant observation of important places for young people to hang out, Dunkley (2004) interviewed nine young people living in northern Vermont to explore their rural youth geographies. In exploring the 'spatialities of teenagerhood', Dunkley (2004: 561) observed that 'young people living in rural areas may reside miles away from their friends and from the centre of town, complicating access to social events and social spaces, and providing a different set of geographies for teens in rural spaces'.

Dunkley's (2004) mapping exercise found that rural youth in Vermont had a number of places to meet their peers including hunting camps, abandoned buildings, construction sites, homes and fields, many of which were miles apart. These sites, however, were often temporary as the young people were forced out and asked to move on by adults. Furthermore, the changing weather resulted in meeting places changing throughout the year meaning that the young people had to respond creatively to the spaces available to them. The distance between the young people meant that many spent a lot of time travelling to meet each other with the average distance travelled on a Saturday evening being 27 miles. The youth culture was therefore car-dependent.

The consumption of alcohol was a key social activity for the young people consulted, with drinking being mentioned most persistently by young men, highlighting the heavily gendered nature of the rural experience. The young men tended to recollect exciting stories about partying, getting drunk and coming into contact with the police, whilst the young women recounted the frustration they felt about the limitations placed on their social activities by their parents. The sense was that boys needed space to be boys whilst girls needed to be protected from them. All of the young women therefore had to develop strategies to manage their social and spatial exclusion. Some girls complied with their parents' rules; others would venture outside the rules (sometimes only occasionally) keeping this private from their parents. Although equal numbers of young men and women had their driving licence, girls were less likely to have access to a car, thereby giving young men greater spatial mobility and opportunities.

Focusing upon the border between Vermont and Quebec, Dunkley (2009) highlights how it identifies the young people as American, suppresses their status as minors and offers them the opportunity to access an adult world due to the drinking age being 18 in Quebec rather than 21 as it is in Vermont. In order to visit bars in Quebec, however, aside from transport, the relevant social networks are required, and so only certain young people engage in such trips. Driving can also be dangerous due to weather conditions and the trip can take up to five hours, meaning that drivers need to be alert for a long period of time in order to remain safe.

Having a section about the consumption of alcohol in a book about young people may work to reinforce stereotypes about young people and binge drinking, yet there are many young people who – for a range of reasons – choose not to drink. Nairn *et al.* (2006) conducted research in urban New Zealand, focusing specifically upon young people who do not drink. They selected what they call the 'school formal' (Nairn *et al.* 2006: 289), the school ball or prom, as the site for their work as this is a context in which alcohol is unlikely to be available. Working with young people from two high schools, and using individual and group interviews, they found four ways in which young people constituted themselves as people who do not drink.

First, some young people justified and legitimated their lack of participation in drinking alcohol through emphasising the importance of not drinking for sport or health-related reasons, whilst others drew upon religious or cultural arguments. A second set of strategies involved young people challenging the assumption that drinking is fun by drawing upon a range of ways in which they can be cool and have fun with their friends without consuming alcohol. The third approach involved young people arguing that drinking too much alcohol was humiliating and degrading, and that drunken bodies were abject. Some of the young people adopted this position having been the named driver on a night out, witnessing the changing behaviours of their friends or through being forced to act as a carer for a drunken peer. The fourth approach involved young people passing as drinkers. Here, young people may have one alcoholic drink that lasts all night; others may have only non-alcoholic drinks but satisfy the people they are with by engaging in drinking of some form; and others still may act intoxicated even though they are not.

Working places

By focusing upon young people's leisure and drinking places, this chapter has tended to focus on those aspects of young people's urban and rural lives that are often viewed negatively and reinforced through stereotypes. I turn now to focus upon young people and the place of work in urban and rural areas. As White and Wyn (2008) note, in order to understand young people's positions as workers, it important to understand the place of work and to appreciate the ways in which work is changing in society. In many ways, gaining an independent income is a key aspect of young people's transitions to increased autonomy and adulthood. However, although earning money is important, the type of work available also has important consequences for young people's identities, well-being and self-esteem.

One approach to understanding work is to recognise the different spheres of economic activity available to young people. White and Wyn (2008: 183–4) describe five main spheres of economic activity:

- the formal waged sphere, which includes paid employment and which is taxed and state regulated;

- the informal waged sphere, which includes paid employment on a cash-in-hand basis, which means it is off the books in terms of formal state regulation of work and taxes;
- the informal non-waged sphere, which includes goods and services being exchanged without monetary payment, as in the case of domestic labour within the parental home;
- the welfare sphere, which includes income benefits received from the state to assist young people in education, training and while unemployed;
- the criminal sphere, which includes activities that cannot be undertaken legally in any economic sectors, on or off the books.

A key economic and social change in recent years has been the increasingly insecure nature of work. White and Wyn (2008) note that precariousness is now a characteristic of many different sectors of the labour market and not only those involving young people. As Katz (1998, 2004) highlights, global economic restructuring has implications for young people in a diverse range of settings, not only across urban and rural contexts, but across nations and continents.

Shucksmith (2004) notes that young people from rural areas in the UK tend to be drawn into one of two labour markets: the national labour market, which is characterised as being distant, relatively well-paid and having career opportunities, and the local labour market, which tends to feature poorly paid, insecure, unrewarding work with fewer career development opportunities. Particular identities, often connected with social class, and qualifications are often needed for young people to access urban or national labour markets. Furthermore, in order to access work, people working in rural areas need their own car or access to transport at the very least, yet low wages and a shortage of affordable housing often restrict access to car ownership.

In the Australian context, Gibson and Argent (2008) observe that levels of rural youth out-migration have increased over the past decade leading to concerns from those observing these social changes about the future sustainability of their villages and communities. Although assumptions are often made about the processes and motivations inherent within these changes in youth migration, Gibson and Argent (2008) suggest a lack of qualitative research means that academics have a relatively poor understanding of the subtlety of rural youth migration, especially since most work within demography and population geography is dominated by quantitative studies. As they observe, 'migration is an inherently complex spatial and social phenomenon, yet only a relatively small portion of that complexity is captured by census statistics' (Gibson and Argent 2008: 136).

Urban labour markets in many countries have changed dramatically over the last couple of decades with transformations including the move from largely manufacturing work to service sector employment, increasing numbers of women entering the workforce, more part-time work and the increasingly insecure nature of work (McDowell 2003). A key change has been the rapid expansion of the service sector, which has become increasingly polarised. On one hand, there are the jobs at the top end of the sector, which includes lawyers, accountants, business managers and bankers. At the opposite end of the sector, the jobs available are more likely to involve servicing the economy through working in retailing, catering, bar work or coffee shops (Image 12.1). As McDowell (2005: 22) observes:

> What unites both types of worker . . . is the significance of bodily performance, in which the personal attributes of an employee are a key part of the service to be exchanged, whether it is financial advice of the most arcane nature or the purchase of a cappuccino in a city centre coffee shop.

One of the key challenges this poses for young people is the demands it places upon them in terms of their skills as well as their bodily performance and management of their identities. Drawing upon her work in merchant banks in the city of London, McDowell (2005) has explored the types of masculinities evident within this workplace environment. The highly competitive and high-risk work environment is clearly suited to young men who present particular forms of masculinity associated with being loud, commanding and aggressively heterosexual. Across the road from the merchant banks in the board rooms of the corporate financial sector – what McDowell (2005: 24) calls 'the other' side of merchant banking' – such behaviour would be inappropriate:

> Here concerns with weight, scent and clothes, the production of a particular version of the dominant and most highly valorised masculinity characterised by a tight, trim, white, middle-class-body appropriately garbed in dark suits and crisp white shirts, is what is required by the employer.
>
> (McDowell 2005: 25)

Image 12.1 At work.

Clearly these issues are particularly pertinent for educated young men (and increasingly young women) who choose to enter such forms of work. Such demands also work to exclude and isolate particular groups of young people from entering such forms of employment not only on the grounds of educational qualifications but also on account of bodily presentation and deportment.

The other end of the job market offers further issues for young people and young men in particular. Jobs in the lower end of the service sector – such as working in retail outlets, coffee shops and McDonalds – present a similar set of challenges in that employees are looking for young people with the ability to serve customers in a polite, efficient and courteous manner. In such jobs, employers often prefer young women who are stereotyped as having more appropriate social skills and abilities to serve customers. The young men McDowell (2005) interviewed in Cambridge and Sheffield in the UK were often forced to take up jobs in the service sector, as this was often the only type of work available. This work nearly always involved these working-class young men working hard, displaying deference to customers and working early or late shifts. Although both young men and young women were asked to perform similar types of work in such employment, there was also a strong sense that they were both disposable and could be replaced at any time.

Ethical and methodological considerations

Read the following two articles and detail the main ethical and methodological issues involved in doing research with young people in urban and rural settings. What are the ethical issues that are consistent in both contexts? What about ethical considerations that are more prevalent in rural contexts compared with urban contexts and vice versa? What influence do the multiple identities of young people have on these considerations?

Leyshon, Michael (2002) On being 'in the field': practice, progress and problems in researching young people in rural areas. *Journal of Rural Studies*, 18(2) 179–91.

McDowell, Linda (2001) 'It's that Linda again': ethical, practical and political issues involved in longitudinal research with young men. *Ethics, Place and Environment* 4(2) 87–100.

Key themes

- Both rural and urban contexts have strong stereotypes associated with them that influence perceptions and attitudes about the presence of young people in both locations. Young people's identities are shaped by these stereotypes in diverse ways, yet there is also much evidence of young people recreating, resisting and reworking these dominant discourses in creative ways. Young people growing up in both urban and rural locations often have to share space with adults and often conflict with both adults and other groups of young people over the use of such spaces.

- Lack of investment in places for young people in cities and the increasing privatisation of the countryside are continuously restricting the places available to young people in both contexts. Despite this, research clearly shows young people's agency creating leisure places in urban and rural contexts.
- The experience of nightlife spaces can usefully be considered through critically appraising the processes of production, regulation and consumption in operation.
- Participating in drinking cultures is an important aspect of young people's lives in urban and rural contexts, although issues of access, mobility and anonymity often differentiate urban experiences from rural.
- There has been a sharp decline in the manufacturing sector in recent years along with a rise in the service sector, with the nature of work available to young people becoming increasingly insecure and precarious.

Project ideas

What are some of the ethical and methodological challenges of researching young people in rural or urban areas?

Critically discuss the similarities and differences between the experiences of urban and rural youth.

What are the similarities and differences between urban and rural youth in terms of their use of leisure places?

What are the economic and social challenges faced by young people looking for work in contemporary society?

Suggested further reading

This excellent book explores the complexity of the night-time economy and the structures and processes in operation at such spaces:

Chatterton, Paul and Hollands, Robert (2003) *Urban nightscapes: youth cultures, pleasure spaces and corporate power*. London: Routledge.

This is a useful account of the ways in which young men in rural areas construct and contest their masculine identities:

Kraack, Anna and Kenway, Jane (2002) Place, time and stigmatised youthful identities: bad boys in paradise. *Journal of Rural Studies* 18(2) 145–55.

This is an excellent exploration of the identities and practices of young people in rural areas:

Leyshon, Michael (2008) The between-ness of being a rural youth: inclusive and exclusive lifestyles. *Social and Cultural Geography* 9(1) 1–26.

This is a useful account of the ways in which young men's access to employment in cities has been shaped by recent economic changes:

McDowell, Linda (2003) Masculine identities and low-paid work: young men in urban labour markets. *International Journal of Urban and Regional Research* 27(4) 828–48.

This article is an excellent exploration of the intersections of age, race, gender and fear for people living in urban areas:

Pain, Rachel (2001) Gender, race, age and fear in the city. *Urban Studies* 5–6: 899–913.

This article charts the complexities of young people's lives in rural areas and sets out an agenda for future research:

Panelli, Ruth (2002) Young rural lives: strategies beyond diversity. *Journal of Rural Studies* 18(2) 113–22.

This excellent article is one of the few to explore young people's experiences in the context of the urban–rural continuum:

Vanderbeck, Robert and Dunkley, Cheryl Morse (2003) Young people's narratives of rural–urban difference. *Children's Geographies* 1(2) 241–59.

13 Conclusions

Young people, place and identity

In this book, I have tried to highlight something of the breadth of young people's lives. We have seen that, although young people may be similar in age, their experiences are very diverse depending on what other identities they possess and where they live. The various chapters have charted the complexities of what it means to be young, the details of young people's engagements with different scales, themes and sites, and the multiple challenges young people face on a daily basis. What it means to be a young person has transformed rapidly over the last few decades as broad changes in the economy, the availability of work and access to education (see, for example Chapters 8, 9 and 12) have altered how young people manage the transition to adulthood and negotiate their youth. Furthermore, this book has also explored the ways that young people are frequently misunderstood and misrepresented, as ageism works to stigmatise their behaviour, associate them with negative activities and demonise their everyday practices. At the same time, young people are not simply victims of these misunderstandings and instead respond in interesting and creative ways to the ageist processes they are subjected to.

This book has demonstrated that as well as being identified by their age, young people are also identified by a range of other markers which operate to empower and enable or exclude and oppress them. The ways in which young people are classed, sexualised, racialised, disabled and gendered alongside their sexual orientation, their body size and their religious affiliations all combine to shape who young people are, how people respond to them and how they take their place in society. As I explored in the introduction, these identities are relational and so are shaped by senses of similarity and difference (Lawler 2008) as young people may feel similar to other social groups whilst being very different from others. Furthermore, these identities are not simply labels to describe individual qualities and are instead connected to broader processes that make different identities matter in particular ways. These processes include, for example, racism, sexism and religious intolerance, and there are many examples of these processes at work throughout this book. From their embodied practices, their experiences of home and their negotiations of public space, it is clear that a range of processes of disadvantage, exclusion and discrimination are at work in determining young people's engagements with place. Examples here include the territorialism young people manage in their local communities (see Chapter 6) or the homophobia experienced by young gay men at school (see Chapter 9).

Place figures prominently in the accounts provided in this book. Young people possess a range of social identities and are situated within the processes which make these identities matter. Moreover, where young people are, the places they use and the localities they engage with all matter too. Young people's identities change across different places and may be interpreted differently according to where young people are. Again, lots of examples, such as young people's drinking behaviours (see Chapter 12), clubbing cultures (see Chapter 4) and experiences of political turmoil (see Chapter 11) are all evidence of this. Overall then, although there is much research about young people's identities, it is crucial also to consider the ways in which locality matters and to think critically about how place makes a difference as young people form particular identities, resist other social identities and remake others forms of identification.

Interconnecting scales, themes and sites

Although the chapters of this book have explored specific scales, themes and sites, it is important to recognise that these do not exist as individual units or bounded entities. As Marston (2004: 173) states 'scale is not a pre-existing category waiting to be applied, but a way of framing conceptions of reality'. Scale is rather artificial, but it is nonetheless a useful device for understanding the world and appreciating how young people negotiate place and identity. I could have organised this book differently but have focused on scales, themes and sites in order to emphasise the complex ways in which place matters to young people's identities. In this book, each scale, theme or site is a useful platform from which to think about the intersections between the various places that young people engage with as they construct and contest their identities. At the same time, these different scales and sites are produced by social activities and interactions between young people and people of all ages and so there are politics that result in the creation of different scales (Herod and Wright 2002, Smith 1993). For example, there are sets of policies and politics that create the scale of the institution and the social contexts that young people negotiate whilst at school, college, university or in residential care (see Chapter 9). Furthermore, there are power relations at play in society which result in adults being in control of public space (see Chapter 10). This often marginalises young people and results in them having to recreate public spaces for their own purposes.

The scales, themes and sites that are the subject of each chapter are therefore interconnected with each other rather than being separate and discrete. As such, the body, the home or public space do not exist as individual units or bounded entities and instead are interconnected with all other scales and sites. For example, the emos, goths and other youth subcultures discussed in Chapter 4 do not only have particular embodied practices and markings; they also have particular experiences of home spaces and may occupy specific public spaces in their home cities as they hang out with their peers. As such, these young people's identities are not only about their bodies, but also about how they use, claim and/or resist the other scales, themes and sites that they negotiate on a daily basis.

Each scale, theme or site shapes and is shaped by other scales, themes and sites in diverse and complex ways. Furthermore, these processes have material outcomes (Marston 2004) and so young people's lives are influenced by the ways in which different scales, and the relationships between and within them, are produced and experienced. It may be useful to think of these interconnections as a network: 'the network metaphor may help articulate a very different sense of scale and the scale's relationships within which people and places are bound, a sense in which specific places are seen as simultaneously global and local (and regional and national) without being wholly one or the other' (Herod and Wright 2002: 8). A useful example to consider here is Dorling's (2001) account of his journey to school (see Box 1.2) where he discusses the different scales, themes and sites he negotiated whilst walking to school. Furthermore, although these different places – the body, public space, the street, the school – are all networked together, this example also points to the inequalities between, for example, those who walk to school and those who are driven to school. As such, the influence that each scale, theme and site has on each other also has consequences for the social position of young people and their access to particular resources.

It is important, therefore, to think about the ways in which young people operate in and across these scales and sites, and how they produce different contexts in the process. One example of the ways in which young people's lives shape and are shaped by interconnected scales, themes and sites can be found in Kathrin Hörschelmann and Nadine Schafer's (2005, 2007) research with young East Germans (see Chapter 8 and Box 8.2). They note that globalisation is felt, practised and managed in a diverse number of locations and 'is always (re)articulated through the actions of specific individuals in the performance of their identities' (Hörschelmann and Schafer 2005: 238). As such, the global does not exist on its own and is instead experienced by young people in different locations, such as through their experiences of home, their daily embodied practices and behaviours, and what they choose to do during their leisure time. Through their interviews as well as through the use of photographs, Hörschelmann and Schafer (2007) show that their young participants experience and engage with global cultural flows (see Box 8.2) through their everyday practices and processes of identity formation. As such, the young east Germans that they worked with engaged with multiple scales, themes and sites on an everyday basis. Through their bodies, such as through dance or through learning a new language, through the organisation and decoration of their personal spaces and through their engagements with difference in the city, the young people simultaneously connected with global issues whilst also being national, urban and embedded in local social and political relations.

A second example of the interconnectedness of scales, sites and themes can be seen in my own work with young Muslim men in

Scotland (see Chapters 4 and 7). Drawing upon focus group and interview data, this project explored the everyday lives and social identities of the young men and clearly demonstrated their engagements with, and entanglements within, multiple scales, themes and sites. The young men's articulations of their identities demonstrated the ways in which they simultaneously claimed particular identities and struggled to belong within others. Their local lives tended to be organised around the key frameworks of home, school or work, mosque and leisure time (Hopkins 2008a), with their urban lives being locations where their multiple understandings of living in ethnically segregated neighbourhoods were played out (Hopkins and Smith 2008). Furthermore, in this context, the young men also sought to claim attachment to the nation through their affiliation to Scottishness whilst also struggling to belong fully within this as a result of racial harassment experienced locally. Whilst the events of 11 September 2001 worked to heighten their sense of difference, it also encouraged some to withdraw to the private spaces of the home (Hopkins and Smith 2008) as the nature of the street and public space changed in meaning and experience (Hopkins 2007e). Here, then, we see that the young men's lives shape and are shaped by the multiple scales and sites of social life. Moreover, these scales also shape, influence and change each other in response to local issues, regional or national concerns and global events.

So, all of the scales, sites and themes discussed in this book are interconnected as they influence young people's lives, structure where they can go and shape who they are. At the same time, however, young people are not simply the victims of these processes and instead may respond in creative and interesting ways to resist, challenge or overturn them. For example, consider how young people make space for themselves in their local neighbourhoods and communities (see Chapter 6) or the imaginative ways they make and remake public spaces through parkour, driving, car cultures, skateboarding and their use of mobile phone technology (see Chapter 10). Alongside these engagements with particular localities, young people also engage with the multiple processes of identification. Identity formation is relational and relies on senses of similarity and/or difference. It is also shaped by inclusionary and exclusionary processes (see Table 1.1) which work to mark out particular identities as desirable and powerful and others as weak and irrelevant. These processes also change through space and across time.

Multiple frameworks for understanding young people, place and identity

In the introduction, I discussed three concepts that can be useful in understanding age: intergenerationality, intersectionality, and lifecourse. Each of these offers different sets of perspectives and different takes on the experiences of age, and together they may offer a rich set of insights into young people's everyday lives. These concepts are not simply detached academic constructions but can be seen in the ways in which young people negotiate place, construct their identities and experience growing up (Hopkins and Pain 2008, Pain and Hopkins 2009). Again, there are many such examples in this book. Young people's negotiations of intergenerational relations – with parents, grandparents and other adults – can be seen in how they experience home as well as how they leave home (and often return) and how they manage setting up their own homes (see Chapter 5). A useful example of this is provided in Box 5.4 which explores the experiences of young gay men coming out to their heterosexual fathers, the different responses that result and the ways in which this changes understandings and connections to home (Skelton and Valentine 2005).

Alongside the insights offered by adopting an intergenerational perspective, intersectionality offers another perspective on the topic of young people, place and identity that helps us to acknowledge that young people are not only identified by age but possess a complex range of other social identities which shape the type of young people they are. Furthermore, this also helps us to appreciate the social processes of discrimination and disadvantage that influence the inclusions and exclusions that young people experience in different places. There are many examples of the ways intersectionality works in young people's lives in this book. Consider how young people are racialised and classed according to their embodied practices (see Chapter 4), how their national affiliation and experience of global events adds further layers of complexity to their identities (see Chapters 7 and 8) or how their peer-group affiliations, locality and everyday practices shape their experiences of public space (see Chapter 10).

Understanding the lifecourse is the third perspective mentioned in the introduction and highlighted at various points throughout this book. Closely connected with intergenerationality, the lifecourse allows us to see young people as negotiators of different fluid and contested stages or transitions which form the lifecourse. This also helps to see young people and experiences of youth as interconnected to people of other ages as different lifecourse stages may overlap and interrelate. Again, there are lots of examples in the book of the ways in which young people are positioned on the lifecourse and negotiate it in multiple ways. Most obviously, there is the discussion about transitions to adulthood (see Chapter 11) alongside explorations of young people's transitions to independent living (see Chapter 5) and their routes through different educational and economic situations (see Chapters 9 and 12).

The exploration of young people, place and identity in this book has also shown that – in conjunction with using intergenerationality, intersectionality and lifecourse to understand young people's lives – research within youth studies tends to be characterised by one of two approaches. The first approach has its origins in cultural studies and the work of the Chicago School and the Birmingham Centre for Cultural Studies (see Box 1.3) and tends to focus upon the cultural practices of young people within particular subcultural groups (see Chapters 4 and 10). The second approach tends to focus upon the ways in which young people negotiate and manage key transitions (see Box 1.3) and is partly motivated from concerns around the changing nature of the economy and the collapse of the youth labour market in the 1970s and 1980s (see Chapters 5, 9, 11 and 12).

A key issue for youth research and future work about young people, place and identity relates to the ways in which understandings of intergenerationality, intersectionality and lifecourse, alongside youth subcultures and youth transitions, are used and applied to research with young people. As each approach has its own strengths and weaknesses, there is much to be said for using them together, where appropriate, in order to enrich and enhance our understanding and appreciation of what it means to be young. With regard to subcultures and transitions, Rob MacDonald and Tracy Shildrick (2007: 350) have argued that 'there is value – in both traditions of youth research – in attempting to conceptualise the way that issues of youth culture impact on youth transition, and vice versa' (see also Shildrick and MacDonald 2005). As such, future work about young people, place and identity could usefully take up this challenge in order to develop and progress understandings about young people, place and identity.

Fundamental to these transitions, subcultures, lifecourses, intergenerationalities and intersectionalities are the ways in which young people negotiate place, how different spaces shape who young people are, and the ways in which young people make space for themselves and for others. Where young people grow up, the places they encounter and the localities that they negotiate during their everyday lives have significant consequences for who they are, where they can and cannot go, and who they may become. Place matters and geography makes a difference.

Youthful futures

In concluding this book, I consider what the future holds for young people and their negotiations of place and identities. In some respects, young people's futures will be shaped by the structures of the institutions, family, home, school, university etc. they navigate through, and the ways in which these transform or not through time. In many countries, we continue to see the massive expansion in higher education resulting in more young people than ever before studying for many years after leaving school. The future for young people will also be influenced by the changing nature of the economy, the availability of work and associated changes in technology and globalisation. Again, we have already seen how young people have struggled through changes in the availability of different types of work, yet the increasing fragility of the labour market is likely to be a key issue for them for many years to come. Issues such as housing availability and the resilience of household relations are also important factors for young people's futures. Furthermore, future political issues and global events will also have a major influence over young people's negotiations of their everyday lives. We see the influence that political conflict and terrorism has on how young people are identified in exclusionary and unjust ways; however, continuing political unease is likely to have marked consequences for how young people experience growing up in different places and times and the ways in which this influences them.

Alongside learning how to negotiate the changing nature of institutions, the economy and political structures, young people also have to manage the ways in which they are regularly stigmatised and misrepresented by the media and by those in power (Jeffrey and Dyson 2008). We see this in the complex ways in which young people are excluded and marginalised from particular places, are assumed to be causing trouble, or are blamed for their particular educational or economic situations. A key issue here, then, is for young people to be involved and included, for them to be listened to, respected and valued for who they are and who they want to become. Given the structures they have to negotiate and the multiple forms of misrepresentation they negotiate, it is crucial that youth researchers continue to conduct critical, ethical and engaged research that is sensitive to young people's personal circumstances, economic situations and their own values and beliefs (see Chapters 6 and 12). Ideally, such research will be theoretically informed (for example, using the frameworks identified above), empirically grounded (involving direct engagement with young people's voices and experiences) and, where appropriate, will inform and critique policy and practice in order to improve the experiences and circumstances of young people in the future.

Key themes

- The scales, sites and themes explored in this book are interconnected and overlapping and are often experienced simultaneously by young people rather than being separate and discrete.
- There are multiple frameworks for understanding young people's lives. Alongside intergenerationality, intersectionality and lifecourse, youth research tends to be dominated by work about subcultures and transitions.
- Future research about young people, place and identity should ideally be theoretically informed and empirically grounded whilst also being sensitive to the various ethical issues arising in engaging with young people in research.

Project ideas

Discuss the ways in which your experience of place and identity formation is shaped by some of the interconnections of the scales, sites and themes explored in this book

What are the strengths and weaknesses of intergenerationality, intersectionality or lifecourse as frameworks for understanding young people, place and identity?

What do approaches informed by youth subcultures and youth transitions add to our appreciation of young people, place and identity?

What are they key factors determining the futures of young people today? How does *where you are* have a role to play in this?

Suggested further reading

This is a useful collection of essays that explore the multiple meanings of power and scale:

Herod, Andrew and Wright, Melissa W. (eds) (2002) *Geographies of power: placing scale*. Oxford: Blackwell.

This article charts the ways in which young Muslim men struggle to belong in the context of the global, national and local:

Hopkins, Peter (2007) Global events, national politics, local lives: young Muslim men in Scotland. *Environment and Planning A* 39(5) 1119–33.

This excellent account describes the ways in which young people's experiences are global yet locally negotiated:

Hörschelmann, Kathrin and Schafer, Nadine (2005) Performing the global through the local: globalisation and individualisation in the spatial practices of young East Germans. *Children's Geographies* 3(2) 219–42.

This is an excellent account of young people's experiences of growing up in poor places and how this is shaped by broader social and political processes:

MacDonald, Robert and Shildrick, Tracy (2007) Street corner society: leisure careers, youth (sub)culture and social exclusion. *Leisure Studies* 26(3) 339–55.

This overview of geographies of age, maps out approaches to advancing understandings of age and place:

Pain, Rachel and Hopkins, Peter (2009) Social geographies of age and ageism: landscapes, lifecourses and justice. Smith, Susan J., Pain, Rachel, Marston, Sallie and Jones III, John Paul (eds) *Handbook of social geography*. London: Sage.

This useful article charts the strengths of the subcultural approach and what this offers to youth studies:

Shildrick, Tracy and MacDonald, Robert (2005) In defence of subculture: young people, leisure and divisions. *Journal of Youth Studies* 9(2) 125–40.

Appendix A: Key authors

Here is some further information about the key authors who write about young people, place and identity. It may be useful to read some of the work of each author and answer the following questions: What perspective does the author have on the concept of youth? Do they use or propose a specific approach to understanding young people's lives? What research methods do they use and what do they say about research ethics? What scales, sites and themes does their work draw attention to?

Stuart Aitken is Professor of Geography and Director of the Center for Interdisciplinary Studies of Youth and Space at San Diego State University, California, and is a leading international contributor to understandings of young people, place and identity. His work has focused on families and communities, children and youth, and geographies of film, and he is the Commissioning Editor for North America for *Children's Geographies*. He has published a number of books relevant to young people, place and identity including *Geographies of young people: the morally contested spaces of identity* (Routledge) which was published in 2001 and *Family fantasies and community space* (Rutgers University Press), published in 1998.

Aitken, Stuart C. (2001). Schoolyard shootings: racism, sexism and moral panics over teen violence. *Antipode* 33(4) 594–600.

Aitken, Stuart C. (2001). Global crises of childhood: rights, justice and the unchildlike child. *Area* 33(2) 119–27.

Aitken, Stuart C. (2000) Fathering and faltering: 'Sorry, but you don't have the necessary accoutrements' *Environment and Planning A* 32(4) 581–98.

Les Back is Professor of Sociology in the Department of Sociology at Goldsmiths, University of London, where he is also associated with the Centre for Urban and Community Research. His work has made important interventions into understandings of young people's social and spatial constructions and contestations of neighbourhoods and communities. His work has drawn attention to the ways in which gender, popular culture, music and, in particular, race and ethnicity influence young people's everyday lives and social interactions. In 1996, he published the landmark book *New ethnicities and urban culture: racisms and multiculture in young lives* (UCL Press).

Back, Les (2005) 'Home from home': youth, belonging and place. Alexander, Claire and Knowles, Caroline (eds) *Making race matter: bodies, space and identity*. New York: Palgrave Macmillan. 19–41.

Back, Les (2002) The fact of hybridity: youth, ethnicity, and racism. Goldbery, David Theo and Solomos, John (eds) *A companion to racial and ethnic studies*. Oxford: Blackwell. 439–54.

Back, Les and Keith, Michael (1999) 'Rights and wrongs': youth, community and narratives of racial violence. Cohen, Phil (ed.) *New ethnicities, old racisms*. Zed Books: London. 131–62.

Caitlin Cahill is an Assistant Professor in the Department of City and Metropolitan Planning at the University of Utah and is a member of the Participatory Action Research Collective at City University of New York. Caitlin's research is characterised by a focus on engagement, interdisciplinarity and community studies. She is a leading contributor to debates about participatory action research in work about young people, place and identity, having worked collaboratively with a group of young urban womyn of colour in New York (see www.fed-up-honeys.org/) exploring issues of stereotyping, everyday lives, senses of community and constructions of identities. Caitlin co-edited (with Roger Hart) four issues of *Children, Youth and Environments* in 2007 on the theme of 'pushing the boundaries: critical international perspectives on child and youth participation' focusing upon participatory approaches to community research, development and governance in nine regions of the world.

Cahill, Caitlin (2004) Defying gravity: raising consciousness through collective research. *Children's Geographies* 2(2) 273–86.

Cahill, Caitlin and Torre, M.E. (2007) Beyond the journal article: representation, audience, and the presentation of participatory research. Kindon, Sara, Pain,

Rachel and Kesby, Mike (eds) *Connecting people, participation and place: participatory action research approaches and methods*. London: Routledge.
Cahill, Caitlin (2007) The personal is political: developing new subjectivities in a participatory action research process. *Gender, Place, and Culture* 14(3) 267–92.

Roger Hart is Professor of Environmental Psychology at the Graduate Center of the City University of New York, Director of the Children's Environments Research Group and Associate Editor of the journal *Children, Youth and Environments*. He has written numerous books, reports, articles and other publications about young people, place and identity, and is particularly known for his work about children and young people's participation and children's everyday environments. He wrote *Children's experiences of place: a developmental study* (Irvington) in 1978, and in 1997, he published *Children's participation: the theory and practice of involving young citizens in community development and environmental care* (Earthscan).

Hart, Roger (2002) Containing children: some lessons on planning for play from New York City. *Environment and Urbanization* 14(2) 135–48.
Hart, Roger and Rajbhandary, Jasmine (2003) Using participatory methods to further the democratic goals of children's organizations. *New Directions for Evaluation* 98 61–75.
Cahill, Caitlin and Hart, Roger (2006) Pushing the boundaries: critical international perspectives on child and youth participation series introduction. *Children, Youth and Environments* 16(2) 1–3.

Robert Hollands is Professor of Sociology in the School of Geography, Politics and Sociology at Newcastle University. He is a leading contributor to a number of debates about youth studies and sociologies of youth having published widely on this topic. In particular, he has made important contributions to understandings of young people's place within the night-time economy and young people's negotiation of these places within the urban context. In 1990, he published *The long transition: class, culture and youth training* (Palgrave Macmillan), and in 2003, he co-authored (with Paul Chatterton) *Urban nightscapes: youth cultures, pleasure spaces and corporate power* (Routledge).

Chatterton, Paul and Hollands, Robert (2002) Theorising urban playscapes: producing, regulating and consuming youthful nightlife city spaces. *Urban Studies* 39(1) 95–116.
Hollands, Robert G. (2002) Divisions in the dark? Youth cultures, transitions and segmented consumption in the night-time economy. *Journal of Youth Studies* 5(2) 153–73.

Hollands, Robert (2004) 'Rappin' on the reservation: Canadian Mohawk youth's hybrid cultural identities. *Sociological Research Online* 9(3). Available at www.socresonline.org.uk/9/3/hollands.html.

Kathrin Hörschelmann is a Lecturer in the Department of Geography and Associate Director of the Centre for the Study of Cities and Regions at the University of Durham. She is a cultural and political geographer and her work has explored the complexities of young people's lives in the former east Germany as well as in the UK, exploring issues relating to gendered identities, post-socialism and youth cultures. In particular, she has made important contributions to understanding the ways in which economic change and political issues shape and are shaped by young people's articulations of local, national and global issues. In 2005, she co-edited (with Bettina van Hoven) *Spaces of masculinities* (Routledge), and in 2010, she published an edited collection (with Rachel Colls) on *Contested bodies of childhood and youth* (Palgrave).

Hörschelmann, Kathrin (2008) Youth and the geopolitics of risk after 11 September 2001. Pain, Rachel and Smith, Susan J. (eds) *Fear: critical geopolitics and everyday life*. Ashgate: Aldershot. 139–52.

Hörschelmann, Kathrin, and Schäfer, Nadine (2007) 'Berlin is not a foreign country, stupid!' – growing up 'global' in eastern Germany. *Environment and Planning A* 39(8) 1855–72.

Hörschelmann, Kathrin and Schäfer, Nadine (2005) Performing the global through the local – young people's practices of identity formation in former east Germany. *Children's Geographies* 3(2) 219–42.

Gill Jones is Emeritus Professor of Sociology at Keele University. She is a leading contributor to scholarship about young people's experiences of leaving home and their transitions to adulthood whilst also contributing actively to policy and practice about young people in the UK. She is the author of *Youth, family and citizenship* (Open University Press 1992) with Claire Wallace, *Leaving home* (Open University Press 1995), *The youth divide: diverging paths to adulthood* (Joseph Rowntree Foundation 2002) and *Youth* (Polity 2009). Her work offers critical perspectives on a range of topics including household formation, young people and family life, homelessness, rural out-migration, intergenerational relations and young people and social policy.

Jones, Gill (2000) Trail-blazers and path-followers: social reproduction and geographical mobility in youth. Arber, Sara and Attias-Donfut, Claudine (eds) *The myth of generational conflict*. London: Routledge: 154–73.

Jones, Gill (2001) Fitting homes? Young people's housing and household strategies in rural Scotland. *Journal of Youth Studies* 4(1) 41–62.
Jones, Gill (2002) Experimenting with households and inventing 'home': *International Social Science Journal* 52(164) 183–94.

Cindi Katz is Professor of Geography in Environment Psychology and Women's Studies at the Graduate Center of the City University of New York. Her work focuses upon issues of social reproduction, children's environments and global economic change. In particular, she has made a series of important contributions to understandings about young people, place and identity through her research about the influence that global economic restructuring has on the everyday lives of children and young people. In 1993, she published *Full circles: geographies of gender over the lifecourse* (Routledge) with Jan Monk, and in 2004 she published *Growing up global: economic restructuring and children's everyday lives* (University of Minnesota Press) and *Life's work: geographies of social reproduction* (Blackwell) with Sallie Marston and Katharyne Mitchell.

Katz, Cindi (2008) Childhood as spectacle: relays of anxiety and the reconfiguration of the child. *Cultural Geographies* 15(1) 5–17.
Katz, Cindi (2001) Vagabond capitalism and the necessity of social reproduction. *Antipode* 33(4) 708–27.
Katz, Cindi (1994) Textures of global change: eroding ecologies of childhood in New York and Sudan. *Childhood* 2(1–2) 103–10.

Jane Kenway is a Professor in Education at Monash University, Melbourne, Australia. Her work has made a series of important contributions to understandings about young people, place and identity, especially given her interests in educational research, globalisation and gendered identities. She co-authored *Answering back: boys, girls and feminism in schools* (Routledge) with Sue Willis in 1998, and in 2006 she published *Masculinity beyond the metropolis* (Palgrave Macmillan) with Anna Kraack and Anna Hickey-Moody.

Kenway, Jane (1997) Taking stock of gender reform policies for Australian schools: past, present and future. *British Educational Research Journal* 23(3) 329–44.
Kraack, Anna and Kenway, Jane (2002) Place, time and stigmatised youthful identities: bad boys in paradise. *Journal of Rural Studies* 18(2) 145–55.
Kenway, Jane and Hickey-Moody, Anna (2009) Spatialized leisure-pleasures and masculine distinctions. *Social and Cultural Geography*.

Michael Leyshon is a Senior Lecturer in Human Geography in the Department of Geography, University of Exeter. He is a social and cultural geographer with interests in youth culture, identities and the multiple ways in which young people are excluded from society. He has made a number of important contributions to understanding the complexities of the everyday life of young people living in rural areas in the UK. His work is characterised by its bottom-up, participatory flavour which seeks to give voice to young people, prioritising their needs and experiences, whilst combining this with theoretical and methodological insights from across social and cultural geographies.

Leyshon, Michael (2002) On being 'in the field': practice, progress and problems in researching young people in rural areas. *Journal of Rural Studies* 18 179–91.

Leyshon, Michael (2008) 'We're stuck in the corner': young women, embodiment and drinking in the countryside. *Drugs: Education, Prevention and Policy* 15(3) 267–89.

Leyshon, Michael (2008) The betweenness of being a rural youth: inclusive and exclusive lifestyles. *Social and Cultural Geography* 9(1) 1–26.

Robert MacDonald is Professor of Sociology in the School of Social Sciences and Law at the University of Teesside. He has made a significant contribution to understanding the relationships between young people, place and identity, particularly through his work about young people growing up and experiencing poor neighbourhoods and communities, and has published numerous books, edited collections and articles. His research interests are youth transitions, subcultures and social exclusion, youth crime, drug use and social class and the use of qualitative methodologies. In 1997, he edited *Youth, the 'underclass' and social exclusion* (Routledge), and in 2005 he co-authored (with Jane Marsh) *Disconnected youth? Growing up in Britain's poor neighbourhoods* (Palgrave).

MacDonald, Robert and Shildrick, Tracy (2007) Street corner society: leisure careers, youth (sub)sultures and social exclusion. *Leisure Studies* 26(3) 339–55.

MacDonald, Robert (2006) Social exclusion, youth transitions and criminal careers: five critical reflections on risk in Australian and New Zealand. *Journal of Criminology* 39(3) 371–83.

MacDonald, Robert and Marsh, Jane (2004) Missing school. *Youth and Society* 36(2) 143–62.

Hugh Matthews is Professor, Director of the Centre for Children and Youth, Dean of the Graduate School and Director of Research and Knowledge Transfer at the University of Northampton. He is the inaugural editor of the journal *Children's Geographies* and has been one of the key contributors to debates about the importance of place and locality in the lives of children and young people in a variety of settings, particularly in England. In 1988, he published a co-edited volume (with Liz Bondi) on *Education and society: studies in the politics, sociology and geography of education* (Routledge), and in 1992 he published *Making sense of place: children's understanding of large-scale environments* (Harvester Wheatsheaf).

Matthews, Hugh (2001) Power games and moral territories: ethical dilemmas when working with children and young people. *Ethics, Place and Environment* 4(2) 117–78.

Matthews, Hugh (2001) Participatory structures and the youth of today: engaging those who are hardest to reach. *Ethics, Place and Environment* 4(2) 153–9.

Matthews, Hugh (2003) The street as a liminal space: the barbed spaces of childhood. Christensen, Pia and O'Brien, Margaret (eds) *Children in the city: home, neighbourhood and community*. London: Routledge. 101–17

Anoop Nayak is a Reader in Social and Cultural Geography in the School of Geography, Politics and Sociology at Newcastle University. He has made a series of important contributions to understandings about young people, place and identity, particularly through his ethnographic work about race, ethnicity and gender in the social and cultural lives of young people. He is the author of *Race, place and globalization: youth cultures in a changing world* (Berg) published in 2003, and in 2008 he co-authored (with Mary Jane Kehily) *Gender, youth and culture: young masculinities and femininities* (Palgrave Macmillan).

Nayak, Anoop (2003) Last of the 'real geordies'? White masculinities and the subcultural response to de-industrialization. *Environment and Planning D: Society and* Space 21(1) 7–25.

Nayak, Anoop (2003) Boyz to men: masculinities, schooling and labour transitions in de-industrial times. *Educational Review* 55(2) 147–59.

Nayak, Anoop (2006) Displaced masculinities: chavs, youth and class in the post-industrial city. *Sociology* 40(5) 813–31.

Rachel Pain is a Reader in the Department of Geography at the University of Durham. She has made a number of important contributions to debates within contemporary social geography and, in particular, she is well known for her work about fear of crime, young people's geographies and participatory research. She is one of the first geographers to consider the importance of intergenerational relations in research and has been a key voice in promoting the significance of age within research in geography. She is co-editor of *Introducing social geographies* (Arnold 2001) with Mike Barke, Jamie Gough, Rob MacFarlane, Graham Mowl and Duncan Fuller, co-editor of *Participatory action research approaches and methods: connecting people, participation and place* (Routledge 2007) with Sara Kindon and Michael Kesby and co-editor of the *Handbook of social geographies* (Sage 2009) with Susan Smith, John Paul Jones III and Sallie Marston.

Pain, Rachel (2006) Paranoid parenting? Rematerializing risk and fear for children. *Social and Cultural Geography* 7(2) 221–43.

Pain, Rachel (2003) Youth, age and the representation of fear. *Capital and Class* 60 151–71.

Pain, Rachel (2001) Gender, race, age and fear in the city. *Urban Studies* 38(5–6) 899–913.

Susan Ruddick is Associate Professor in the Department of Geography and Program in Planning at the University of Toronto, Canada. She is the author of *Young and homeless in Hollywood: mapping social identities* (Routledge 1996) and has made important contributions to understanding the experiences of homeless young people, theorising about children's rights and critical understandings of the intersections of social identities.

Ruddick, Susan (1998) Modernism and resistance: how 'homeless' youth sub-cultures make a difference. Skelton, Tracey and Valentine, Gill (eds) *Cool places: geographies of youth cultures*. London: Routledge. 344–62.

Ruddick, Susan (2003) The politics of aging: globalization and the restructuring of youth and childhood. *Antipode* 35(2) 334–62.

Ruddick, Susan (2007) At the horizons of the subject: neo-liberalism, neo-conservatism and the rights of the child Part Two: parent, caregiver, state. *Gender, Place and Culture* 14(6) 627–40.

Tracey Skelton is Associate Professor in the Department of Geography at the National University of Singapore and Professor of Human Geography at Loughborough University. She is arguably one of the key contributors to research about young people, place and identity having established an international reputation for her work in this area. In 1998 she co-edited (with Gill Valentine) *Cool places: geographies of youth cultures* (Routledge), a landmark collection of essays about the contested spaces and places of young people's lives. Her work offers a series of critical insights to the complex ways that power relations and identities – gender, disability, age and race in particular – are salient in shaping, and being shaped by, young people's everyday use of different places. She has also made important contributions to ethical and methodological debates about doing research with young people. Her research includes work with diverse groups of young people such as deaf youth, young women and lesbian and gay youth.

Skelton, Tracey (2000) Nothing to do, nowhere to go? Teenage girls and 'public' space in the Rhondda Valleys, South Wales. Holloway, Sarah and Valentine, Gill (eds) *Children's geographies: playing, living, learning*. London: Routledge.

Skelton, Tracey (2008) Research with children and young people: exploring the tensions between ethics, competence and participation. *Children's Geographies* 6(1) 21–36.

Skelton, Tracey (2009) Children's geographies/geographies of children: play, work, mobilities and migration. *Geography Compass* 3(4) 1430–48.

Gill Valentine is Professor of Geography and Director of the Leeds Social Science Institute at the University of Leeds. A leading international contributor to a variety of topics within contemporary social and cultural geography, she has been one of the leading scholars within the development of studies about children and young people within geography. She co-edited (with Tracey Skelton) the landmark collection *Cool places: geographies of youth cultures* (Routledge) in 1998, and in 2000 she co-edited (with Sarah Holloway) *Children's geographies: living, playing, learning* (Routledge). In 2003, she co-authored (with Sarah Holloway) *CyberKids: children and the information age* (Routledge), and in 2004 she wrote *Public space and the culture of childhood* (Ashgate).

Valentine, Gill (2003) Boundary crossings: transitions from childhood to adulthood. *Children's Geographies* 1(1) 37–52.

Valentine, Gill (2000) Exploring children's and young people's narratives of identity. *Geoforum* 31 257–67.

Valentine, Gill (1999) Being seen and heard? The ethical complexities of working with children and young people at home and at school. *Ethics, Place and Environment* 2(2) 141–55.

Appendix B: Journals about young people, place and identity

Some useful academic journals about young people, place and identity:

For articles about various aspects of young people's lives see:

British Journal of Sociology of Education
Child and Family Social Work
Childhood
Childhoods Today
Children's Geographies
Children, Youth and Environments
Feminism and Psychology
Gender and Education
Journal of Adolescence
Journal of Adolescent Research
Journal of Education and work
Journal of Urban Youth Culture
Journal of Youth and Adolescence
Journal of Youth Studies
Urban Education
Young: Nordic Journal of Youth Research
Youth and Society
Youth Justice: an international Journal
Youth Studies Australia

For articles that draw attention to the significance of place see:

Annals of the Association of American Geographers
Area
Environment and Planning A
Environment and Planning D: Society and Space
Health and Place
Journal of Rural Studies
Population, Space and Place
Progress in Human Geography
Social and Cultural Geography
Transactions of the Institute of British Geographers
Urban Geography
Urban Studies

For articles that draw attention to identities see:

Identities – *Sociological Research Online, Sociology*
Age – *Age and Ageing, Ageing and Society*
Gender – *Gender and Education, Gender, Place and Culture, Women's Studies International Forum*
Race and ethnicity – *Ethnic and Racial Studies, Ethnicities, Migrants and Minorities, Race, Ethnicity and Education*
Disability – *Disability and Society*
Sexuality – *Sexualities*

Appendix C: Research centres and organisations

Research centres and organisations that conduct research connected with young people, place and identity include:

Australian Clearinghouse for Youth Studies
www.acys.info/

Center for Interdisciplinary Studies of Youth and Space, San Diego State University, California, USA
http://geography.sdsu.edu/Research/Projects/ISYS/index.html

Center for Youth Development and Policy Research, Washington DC
http://cydpr.aed.org/

Centre for Children and Young People, Southern Cross University, Lismore, NSW, Australia
http://ccyp.scu.edu.au/index.php

Centre for Children and Youth, University of Northampton, UK
http://www2.northampton.ac.uk/socialsciences/sshome/c-c-y

Centre for Research on Youth at Risk, St Thomas University, Canada
www.stthomasu.ca/research/youth/index.htm

Centre for the Study of Childhood and Youth, University of Sheffield, UK
www.cscy.group.shef.ac.uk/index.htm

Child and Youth Interdisciplinary Research Centre
http://misprd.uow.edu.au/ris_public/WebObjects/RISPublic.woa/wa/default?group = 24

Children's Environments Research Group, City University New York
http://web.gc.cuny.edu/che/cerg/about_cerg/index.htm

Children's Issues Centre, University of Otago, Dunedin, New Zealand
www.otago.ac.nz/cic/

European Youth Observatory
www.diba.es/eyo/

International Clearinghouse on Children, Youth and Media, University of Gothenburg, Sweden
www.nordicom.gu.se/clearinghouse.php

Youth Research Centre, University of Melbourne, Australia
www.edfac.unimelb.edu.au/yrc/

Bibliography

Aitken, Stuart C. (2001a) *Geographies of young people: the morally contested space of identity*. London: Routledge.

—— (2001b) Global crises of childhood: rights, justice and the unchildlike child. *Area* 33(2) 119–27.

Alderson, Priscilla and Morrow, Virginia (2004) *Ethics, social research and consulting with children and young people*. London: Barnardos.

Alexander, Catherine (2008) Safety, fear and belonging: the everyday realities of civic identity formation in Fenham, Newcastle Upon Tyne. *ACME* 7(2) 173–98.

—— (2010) Deviant femininities: the everyday making and unmaking of 'criminal' youth. In Colls, Rachel and Hörschelmann, Kathrin (eds) *Contested bodies of childhood and youth*. Basingstoke: Palgrave. 68–83.

Alexander, Claire (2004) Imagining the Asian gang: ethnicity, masculinity and youth after 'the riots'. *Critical Social Policy* 24(4) 526–49.

Allen, Louisa (2005) Managing masculinity: young men's identity work in focus groups. *Qualitative Research* 5(1) 35–57.

Anderson, Benedict (1983) *Imagined communities*. London: Verso.

Ansell, Nicola (2005) *Children, youth and development*. London: Routledge.

—— (2008) Third World gap year projects: youth transitions and the mediation of risk. *Environment and Planning D: Society and Space* 26 218–40.

Anthias, Floya (2001) New hybridities, old concepts: the limits of 'culture'. *Ethnic and Racial Studies* 24(4) 619–41.

Appadurai, Arjun (1996) *Modernity at large: cultural dimensions of globalisation*. Minneapolis: University of Minnesota Press.

Aries, Philippe (1960) *Centuries of childhood*. New York: Vintage Books.

Arnett, Jeffrey Jensen (2000) Emerging adulthood: a theory of development from the late teens through the twenties. *American Psychologist* 55(5) 469–80.

—— (2001) *Adolescence and emerging adulthood: a cultural approach*. New Jersey: Prentice Hall.

—— (2006) Emerging adulthood in Europe: a response to Bynner. *Journal of Youth Studies* 9(1) 111–23.

Askins, Kye (2007) Codes, committees and other such conundrums. *ACME: An International E-journal for Critical Geographers* 6(3) 350–9.

Back, Les (1996) *New ethnicities and urban culture: racisms and multiculture in young lives*. London: UCL Press.

—— (2000) 'Pale shadows': racisms, masculinity and multiculture. In Roche, Jeremy and Tucker, Stanley (eds) *Youth in society*. London: Sage. 35–48.

—— (2002) The fact of hybridity: youth, ethnicity, and racism. In Goldbery, David Theo and Solomos, John (eds) *A companion to racial and ethnic studies*. Oxford: Blackwell. 439–54.

—— (2005) 'Home from home': youth, belonging and place. In Alexander, Claire and Knowles, Caroline (eds) *Making race matter: bodies, space and identity*. New York: Palgrave Macmillan. 19–41.

Bailey, Brian (2006) Emo music and youth culture. In Steinberg, Shirley R., Parmar, Priya and Richard, Birgit (eds) *Contemporary youth culture: an international encyclopaedia*. Westport, CT: Greenwood Press. 338–42.

Bailey, Cathy, White, Catherine and Pain, Rachel (1999a) Evaluating qualitative research: dealing with tensions between 'science' and 'creativity'. *Area* 31(2) 169–83.

—— (1999b) Response. *Area* 31(2) 182–3.

Barker, John and Weller, Susie (2003) 'Never work with children?' The geography of methodological issues in research with children. *Qualitative Research*.

Barker, John, Kraftl, Peter, Horton, John and Tucker, Faith (2009) The road less travelled – new directions in children's and young people's mobility. *Mobilities* 4(1) 1–10.

Baxter, Jamie and Eyles, John (1997) Evaluating qualitative research in social geography: establishing 'rigour' in interview analysis. *Transactions of the Institute of British Geographers* 22 505–25.

—— (1999) Prescription for research practice? Grounded theory in qualitative evaluation. *Area* 31(2) 179–81.

Bell, David and Valentine, Gill (1997) *Consuming geographies: we are where we eat*. London: Routledge.

Billig, Michael (1995) *Banal nationalism*. London: Sage.

Blunt, Alison and Dowling, Robyn (2006) *Home*. London: Routledge.

Blunt, Alison and Varley, Ann (2004) Geographies of home. *Cultural Geographies* 11(1) 3–6.

Borden, Iain (2001) *Skateboarding, space and the city: architecture and the body*. Oxford: Berg.

Boyden, Jo and Hart, Jason (2007) The statelessness of the world's children. *Children and Society* 21 237–49.

Bruce, Jasmine (2003) Indigenous Youth. Oxfam (eds) *Highly affected, rarely considered*. Youth Commission Report into Globalisation.

Budgeon, Shelley (2003) Identity as an embodied event. *Body and Society* 9(1) 35–55.

Butler, Ruth (1999) The body. In Cloke, Paul, Crang, Philip and Goodwin, Mark (eds) *Introducing human geographies*. London: Arnold. 238–45.

Bynner, John (2005) Rethinking the youth phase of the life-course: the case for emerging adulthood. *Journal of Youth Studies* 8(4) 367–84.

Bytheway, Bill (1995) *Ageism*. Buckingham: Open University Press.

Cahill, Caitlin (2000) Street literacy: urban teenagers' strategies for negotiating their neighbourhood. *Journal of Youth Studies* 3(3) 251–77.

—— (2004) Defying gravity: raising consciousness through collective research. *Children's Geographies* 2(2) 273–86.

—— (2006) 'At risk'? The fed up honeys re-present the gentrification of the Lower East Side. *Women Studies Quarterly* 34(1 and 2) 334–63.

—— (2007a) Negotiating grit and glamour: young women of color and the gentrification of the Lower East Side. *City and Society* 19(2) 202–31.

—— (2007b) Doing research *with* young people: participatory research and the rituals of collective work. *Children's Geographies* 5(3) 297–312.

—— (2007c) Including excluded perspectives in participatory action research. *Design Studies* 28(3) 325–40.

—— (2007d) The personal is political: developing new subjectivities in a participatory action research process. *Gender, Place, and Culture* 14(3) 267–92.

—— (2007e) Repositing ethical commitments: participatory action research as a relational praxis of social change. *ACME: An International E-journal for Critical Geographers* 6(3) 360–73.

Cahill, Caitlin and Katz, Cindi (2008) Young Americans: geographies at the crossroads. *Environment and Planning A* 40 2809–13.

Cahill, Caitlin, Sultana, Farhana and Pain, Rachel (2007) Participatory ethics: politics, practices and institutions. *ACME: An International E-journal for Critical Geographers* 6(3) 304–18.

Casey, Mark (2002) Young gay males' experiences of coming out in the context of school. *Youth and Policy* 75 62–75.

Chandler, Joan, Williams, Malcolm, Maconachie, Moira, Collett, Tracey and Dodgeon, Brian (2004) Living alone: its place in household formation and change. *Sociological Research Online* 9(3). Available on line at http://www.socresonline.org.uk/9/3/chandler.html.

Chatterton, Paul (1999) University students and city centres – the formation of exclusive geographies: the case of Bristol, UK. *Geoforum* 30 117–33.

Chatterton, Paul and Hollands, Robert (2002) Theorising urban playscapes: producing, regulating and consuming youthful nightlife city spaces. *Urban Studies* 39(1) 95–116.

—— (2003) *Urban nightscapes: youth cultures, pleasure spaces and corporate power*. London: Routledge.

Chawla, Louise (2002) Towards better cities for children and youth. In Chawla, Louise (ed.) *Growing up in an urbanising world*. London: Earthscan. 219–42.

Christie, Hazel (2007) Higher education and spatial (im)mobility: non-traditional students and living at home. *Environment and Planning A* 39 2445–63.

Cieslik, Mark (2003) Introduction: contemporary youth research: issues, controversies and dilemmas. In Bennett, Andy, Cieslik, Mark and Miles, Steven (eds) *Researching youth*. New York: Palgrave Macmillan. 1–12.

Clark, William, A.V. and Mulder, Clara, A. (2000) Leaving home and entering the housing market. *Environment and Planning A* 32 1657–71.

Cohen, Stanley (1987) *Folk devils and moral panics: the creation of the mods and rockers*. Oxford: Blackwell.

Coles, Bob, Rugg, Julie and Seavers, Jenny (1999) Young adults living in the parental home: the implications of extended youth transitions for housing and social policy. In Rugg, Julie (ed.) *Young people, housing and social policy*. London: Routledge. 159–81.

Collins, Damian C.A. and Kearns, Robin A. (2001) Under curfew and under siege? Legal geographies of young people. *Geoforum* 32 389–403.

Collins, Jock, Noble, Greg, Poynting, Scott and Tabar, Paul (2000) *Kebabs, kids, cops and crime*. Annandale, NSW: Pluto Press.

Colls, Rachel and Hörschelmann, Kathrin (2009) Editorial: the geographies of children's and young people's bodies. *Children's Geographies* 7(1) 1–6.

—— (eds) (2010) *Contested bodies of childhood and youth*. Basingstoke: Palgrave. 1–21.

Cox, Kevin R. (2002) *Political geography: territory, state and society*. Oxford: Blackwell.

Crenshaw Kimberle (1993) Mapping the margins: intersectionality, identity politics, and violence against women of color. *Stanford Law Review* 43 1241–76.

Crewe, Louise (2001) The besieged body: geographies of retailing and consumption *Progress in Human Geography* 25 629–40.

Crockett, Alasdair and Voas, David (2006) Generations of decline: religious change in 20th-century Britain. *Journal for the Scientific Study of Religion* 45(4) 567–84.

Curtis, Penny (2008) The experiences of young people with obesity in secondary schools: some implications for the healthy school agenda. *Health and Social Care in the Community* 16(4) 410–18.

Curtis, Sarah, Daya, Shari, Khatib, Yasmin, Pain, Rachel, Rothon, Catherine and Stansfield, Stephen (2008) *Mapping links between young people, neighbourhoods, schools and families, with respect to adolescent mental health*. Nuffield Foundation.

Desforges, Luke (2000) Travelling the world: identity and travel biography. *Annals of Tourism Research* 27(4) 926–45.

Donovan, Gregory and Katz, Cindi (2009) Cookie monsters: seeing young people's hacking as creative practice. *Children, Youth and Environments* 19(1) 197–222.

Dorling, Danny (2001) Anecdote is the singular of data. *Environment and Planning A* 33 1335–69.

Dunkley, Cheryl Morse (2004) Risky geographies: teens, gender, and the rural landscape in North America. *Gender, Place and Culture* 11(4) 559–79.

—— (2009) A therapeutic taskscape: theorizing place-marking, discipline and care at a camp of troubled youth. *Health and Place* 15 88–96.

Dunn, Kevin M. (2001) Representations of Islam in the politics of mosque development in Sydney. *Tijdschrist Voor Economische en Sociale Geografie* 92(3) 291–308.

Dunn, Kevin, Klocker, Natasha and Salabay, Tanya (2007) Contemporary racism and Islamophobia in Australia: racializing religion. *Ethnicities* 7(4) 564–89.

Dwyer, Claire, (1998) Contested identities: challenging dominant representations of young British Muslim women. In Skelton, Tracey and Valentine, Gill (eds) *Cool places: geographies of youth cultures*. London: Routledge. 50–64.

Dwyer, Claire (1999a) Veiled meanings: young British Muslim women and the negotiation of differences. *Gender, Place and Culture* 6(1) 5–26.

—— (1999b) Contradictions of community: questions of identity for young British Muslim women. *Environment and Planning A* 31 53–68.

Edmond, Ruth (2003) Putting the care into residential care: the role of young people. *Journal of Social Work* 3(3) 321–37.

—— (2005) Ethnographic research methods with children and young people. In Greene, Sheila and Hogan, Diane (eds) *Researching children's experience: approaches and methods*. London: Sage. 123–40.

Elwood, Sarah (2007) Negotiating participatory ethics in the midst of institutional ethics. *ACME: An International E-journal for Critical Geographers* 6(3) 329–38.

Elwood, Sarah and Martin, Deborah (2000) 'Placing' interviews: location and scales of power in qualitative research. *Professional Geographer* 52(4) 649–67.

Evans, Bethan (2008) Geographies of youth/young people. *Geography Compass* 2(5) 1559–680.

Evans, Helen (2007) Institutions. In Robb, Martin (ed.) *Youth in context: frameworks, settings and encounters*. London: Sage. 185–216.

Fahmy, Eldin (2003) A disconnected generation? Encouraging young people's political participation in the UK. *Youth and Policy* Autumn 1–20.

Falconer, Ryan and Kingham, Simon (2007) 'Driving people crazy': a geography of boy racers in Christchurch, New Zealand. *New Zealand Geographer* 63 181–91.

Ford, Janet (1999) Young adults and owner occupation: a changing goal? In Rugg, Julie (ed.) *Young people, housing and social policy*. London: Routledge. 17–34.

France, Alan (2007) *Understanding youth in late modernity*. Maidenhead: Open University Press.

French, Sally and Swain, John (2004) Young disabled people. In Roche, Jeremy, Tucker, Stanley, Thomson, Rachel and Flynn, Ronny (eds) *Youth in society* (second edition). London: Sage. 199–206.

Furlong, Andy (2009) (eds) *Handbook of youth and young adulthood*. London: Routledge.

Furlong, Andy and Cartmel, Fred (2007) *Young people and social change: new perspectives*. Maidenhead: Open University Press.

Fyfe, Nicholas R. (1998) Introduction: reading the street. In Fyfe, Nicholas R. (ed.) *Images of the street: planning, identity and control in public space*. London: Routledge. 1–12.

Gallagher, Michael (2009a) Data collection and analysis. In Tisdall, Kay, Davis, John and Gallagher, Michael (eds) *Researching with children and young people: research design, methods and analysis*. London: Sage. 65–127.

—— (2009b) Ethics. In Tisdall, Kay, Davis, John and Gallagher, Michael (eds) *Researching with children and young people: research design, methods and analysis*. London: Sage. 11–64.

Gibson, Chris and Argent, Neil (2008) Getting on, getting up and getting out? Broadening perspectives on rural youth migration. *Geographical Research* 46(2) 135–8.

Giddings, Robert and Yarwood, Richard (2005) Growing up, going out and growing out of the countryside: childhood experiences in rural England. *Children's Geographies* 3(1) 101–14.

Gilligan, Robbie (1999) Enhancing the resilience of children and young people in public care by mentoring their talents and interests. *Child and Family Social Work* 4 187–96.

Goheen, Peter G. (1998) Public space and the geography of the modern city. *Progress in Human Geography* 22(4) 479–96.

Greene, Shiela and Hogan, Diane (eds) (2005) *Researching children's experience: approaches and methods*. London: Sage.

Greig, Anne, Taylor, Jayne and MacKay, Tommy (2007) *Doing research with children* (second edition). London: Sage.

Gruffudd, Pyrs (1999) Nationalism. In Cloke, Paul, Crang, Philip and Goodwin, Mark (eds) *Introducing human geographies*. London: Arnold. 199–207.

Grundy, Sue and Jamieson, Lynn (2005) Are we all Europeans now? Local, national and supranational identities of youth adults. *Sociological Research online* 10(3). Available on line at http://www.socresonline.org.uk/10/3/grundy.html.

—— (2007) European identities: from absent-minded citizens to passionate Europeans. *Sociology* 41(4) 663–80.

Hall, Stuart and Jefferson, Tony (eds) (1976) *Resistance through rituals: youth subcultures in post-war Britain*. London: Hutchinson.

Hart, Roger (1997) *Children's participation: the theory and practice of involving young citizens in community development and environmental care*. London: Earthscan.

Heath, Sue (2007) Widening the gap: pre-university gap years and the 'economy of experience'. *British Journal of Sociology of Education* 28(1) 89–103.

Heath, Sue and Kenyon, Liz (2001) Single young professionals and shared household living. *Journal of Youth Studies* 4(1) 83–100.

Heath, Sue, Brooks, Rachel, Cleaver, Elizabeth and Ireland, Eleanor (2009) *Researching young people's lives*. London: Sage.

Hebdige, Dick (1987) *Subculture: the meaning of style*. London: Methuen.

Henderson, Sheila (2007) Neighbourhood. Rob, Martin (ed.) *Youth in context: frameworks, settings and encounters*. London: Sage. 123–54.

Henderson, Sheila, Holland, Janet, McGrellis, Sheens, Sharpe, Sue and Thomson, Rachel (2007) *Inventing adulthoods: a biographical approach to youth transitions*. London: Sage.

Henn, Matt, Weinstein, Mark and Wring, Dominic (2002) A generation apart/youth and political participation in Britain. *British Journal of Politics and International Relations* 4(2) 167–92.

Herod, Andrew and Wright, Melissa W. (2002) Placing scale: an introduction. In Herod, Andrew and Wright, Melissa W. (eds) *Geographies of power: placing scale*. Oxford: Blackwell. 1–14.

Hill, Malcolm (2005) Ethical considerations in researching children's experiences. In Greene, Sheila and Hogan, Diane (eds) *Researching children's experience: approaches and methods*. London: Sage. 61–86.

—— (2006) Children's voices on ways of having a voice. *Childhood* 13(1) 69–89.

Hill, Malcolm and Tisdall, Kay (1997) *Children and society*. Harlow: Pearson.

Hill, Malcolm, Stafford, Anne, Seaman, Peter, Ross, Nicola and Daniel, Brigid (2007) *Parenting and resilience*. York: Joseph Rowntree Foundation.

Hill, Malcolm, Turner, Katrina, Walker, Moira, Stafford, Anne and Seaman, Peter (2006) Children's perspectives on social exclusion and resilience in disadvantaged urban communities. In Tisdall, Kay, David, John, Hill, Malcolm and Prout, Alan (eds) *Children, young people and social inclusion: participation for what?* Bristol: Policy Press. 39–56.

Hinduja, Sameer and Patchin, Justin W. (2008) Personal information of adolescents on the Internet: a quantitative content analysis of MySpace. *Journal of Adolescence* 31 125–46.

Hockey, Jenny and James, Allison (2003) *Social identities across the life course*. New York: Palgrave Macmillan.

Hodkinson, Paul and Lincoln, Sian (2008) Online journals as virtual bedrooms? Young people, identity and personal space. *Young* 16(1) 27–46.

Holdsworth, Clare (2000) Leaving home in Britain and Spain. *European Sociological Review* 16(2) 201–22.

—— (2007) Intergenerational inter-dependencies: mothers and daughters in comparative perspective. *Women's Studies International Forum* 30 59–69.

Holdsworth, Clare and Morgan, David (2005) *Transitions in context: leaving home, independence and adulthood*. Berkshire: Open University Press.

Hollands, Robert (1990) *The long transition: class, culture and youth training*. London: Macmillan.

Holliday, Ruth and Hassard, John (2001) Contested bodies: an introduction. In Holliday, Ruth and Hassard, John (eds) *Contested bodies*. London: Routledge. 1–18.

Holloway, Lewis and Hubbard, Phil (2001) *People and place: the extraordinary geographies of everyday life*. Harlow: Pearson Education.

Holt, Louise (2004) The 'voices' of children: de-centring empowering research relations. *Children's Geographies* 2(1) 13–27.

Hopkins, Peter (2004) Young Muslim men in Scotland: inclusions and exclusions. *Children's Geographies* 2(2) 252–72.

—— (2006) Youth transitions and going to university: the perceptions of students attending a geography summer school access programme. *Area* 38(3) 240–7.

—— (2007a) Global events, national politics, local lives: young Muslim men in Scotland. *Environment and Planning A* 39(5) 1119–33.

—— (2007b) 'Blue squares', 'proper' Muslims and transnational networks: narratives of national and religious identities amongst young Muslim men living in Scotland. *Ethnicities* 7(1) 61–81.

—— (2007c) Thinking critically and creatively about focus groups. *Area* 39(4) 528–35.

—— (2007d) Positionalities and knowledge: negotiating ethics in practice. *ACME: An International E-journal for Critical Geographers* 6(3) 386–94.

—— (2007e) Young Muslim men's experiences of local landscapes after 11 September 2001. In Aitchison, Cara, Hopkins, Peter and Kwan, Mei-Po (eds) *Geographies of Muslim identities: diaspora, gender and belonging*. Aldershot: Ashgate. 189–200.

—— (2008a) *The issue of masculine identities for British Muslims: a social analysis*. Lampeter: Edwin Mellen Press.

—— (2008b) Ethical issues in research with unaccompanied asylum-seeking children. *Children's Geographies* 6(1) 37–48.

Hopkins, Peter and Bell, Nancy (2008) Interdisciplinary perspectives: ethical issues and child research. *Children's Geographies* 6(1) 1–6.

Hopkins, Peter and Hill, Malcolm (2006) *'This is a good place to live and think about the future': the needs and experiences of unaccompanied asylum-seeking children and young people in Scotland.* Glasgow: Scottish Refugee Council.

—— (2008) Pre-flight experiences and migration stories: the accounts of unaccompanied asylum-seeking children. *Children's Geographies* 6(3) 257–68.

Hopkins, Peter and Pain, Rachel (2007) Geographies of age: thinking relationally. *Area* 39(3) 287–94.

—— (2008) Is there more to life? Relationalities in here and out there: a response to Horton and Kraftl. *Area* 40(2) 289–92.

Hopkins, Peter, and Smith, Susan J. (2008) Scaling segregation; racialising fear. In Pain, Rachel and Smith, Susan J. (eds) *Fear: critical geopolitics and everyday life*. Aldershot: Ashgate. 103–16.

Hörschelmann, Kathrin (2008a) Youth and the geopolitics of risk after 11 September 2001. In Pain, Rachel and Smith, Susan J. (eds) *Fear: critical geopolitics and everyday life*. Aldershot: Ashgate. 139–52.

—— (2008b) Populating the landscapes of critical geopolitics – young people's responses to the war in Iraq (2003). *Political Geography* 27(5) 587–609.

Hörschelmann, Kathrin and Colls, Rachel (2010) Introduction: contested bodes of childhood and youth. In Colls, Rachel and Hörschelmann, Kathrin (eds) *Contested bodies of childhood and youth*. Basingstoke: Palgrave. 1–21.

Hörschelmann, Kathrin and Schafer, Nadine (2005) Performing the global through the local – globalisation and individualisation in the spatial practices of young East Germans. *Children's Geographies* 3(2) 219–42.

—— (2007) 'Berlin is not a foreign country, stupid!' – growing up 'global' in eastern Germany. *Environment and Planning A* 39 1855–72.

Horton, John (2001) 'Do you get some funny looks when you tell people what you do?' Muddling through some angst and ethics of (being a male) researching with children. *Ethics, Place and Environment* 4(2) 159–65.

Horton, John and Kraftl, Peter (2008) Reflections on geographies of age: a response to Hopkins and Pain. *Area* 40(2) 284–8.

Hunt, Stephen (2005) *The life course: a sociological introduction.* New York: Palgrave Macmillan.

Hyams, Melissa (2000) 'Pay attention in class . . . [and] don't get pregnant': a discourse of academic success among adolescent Latinas. *Environment and Planning A* 32 635–54.

—— (2002) 'Over there' and 'back then': an odyssey in national subjectivity. *Environment and Planning D: Society and Space* 20 459–76.

—— (2004) Hearing girls' silences: thoughts on the politics and practices of feminist method of group discussion. *Gender, Place and Culture* 11(1) 105–19.

Hyde, Abbey, Howlett, Etaoine, Brady, Dympna and Drennan, Jonathan (2005) The focus group method: insights from focus group interviews on sexual health with adolescents. *Social Sciences and Medicine* 61(12) 2588–99.

Immigration Law Practitioners Assocation (2004) *Working with children and young people subject to immigration control: guidelines for best practice*. London: Immigration Law Practitioner Association.

Ince, Lynda (2004) Young black people leaving care. In Lewis, Vicky, Kellett, Mary, Robinson, Chris, Fraser, Sandy and Ding, Sharon (eds) *The reality of research with children and young people*. London: Sage. 210–26.

Islam, Zoebia (2008) Negotiating identities: the lives of Pakistani and Bangladeshi young disabled people. *Disability and Society* 23(1) 41–52.

Iveson, Kurt (2007) *Publics and the city*. Oxford: Blackwell.

Jackson, Peter and Penrose, Jan (1993) *Constructions of race, place and nation*. Minneapolis: University of Minnesota Press.

Jacobs, Jane M. and Smith, Susan J. (2008) Living room: rematerialising home. *Environment and Planning A* 40 515–19.

James, Allison, Jenks, Chris and Prout, Alan (1998) *Theorizing childhood*. Cambridge: Polity Press.

Jayne, Mark, Holloway, Sarah L. and Valentine, Gill (2006) Drunk and disorderly: alcohol, urban life and public space. *Progress in Human Geography* 30 451–568.

Jeffrey, Craig and Dyson, Jane (2008) *Telling young lives: portraits of global youth*. Philadelphia: Temple University Press.

Jeffrey, Craig and McDowell, Linda (2004) Youth in a comparative perspective: global change, local lives. *Youth and Society* 36(2) 131–42.

Jenkins, Richard (2004) *Social identity*. London: Routledge.

Jones, Andrew (2008) The rise of global work. *Transactions of the Institute of British Geographers* 33(1) 12–26.

Jones, Gill (1995) *Leaving home*. Buckingham: Open University Press.

—— (2002) *The youth divide: diverging paths to adulthood*. York: Joseph Rowntree Foundation.

Jones, Martin, Jones, Rhys and Woods, Michael (2004) *An introduction to political geography: space, place and politics*. London: Routledge.

Karsten, Lia and Pel, Eva (2000) Skateboarders exploring urban public space: ollies, obstacles and conflicts. *Journal of Housing and the Built Environment* 15 327–40.

Katz, Cindi (1998) Disintegrating developments: global economic restructuring and the eroding ecologies of youth. In Skelton, Tracey and Valentine, Gill (eds) *Cool places: geographies of youth culture*. London: Routledge. 130–44.

—— (2004) *Growing up global: economic restructuring and children's everyday lives*. Minneapolis: University of Minnesota Press.

Kearns, Ade and Parkinson, Michael (2001) The significance of neighbourhood. *Urban Studies* 38(12) 2103–10.

Kehily, Mary Jane (2007a) Introduction. Kehily, Mary Jane (ed.) *Understanding youth: perspectives, identities and practices*. London: Sage. 3–8.

—— (2007b) A cultural perspective. Kehily, Mary Jane (ed.) *Understanding youth: perspectives, identities and practices*. Sage: London. 11–44.

Kehily, Mary Jane and Nayak, Anoop (2008) Global femininities: consumption, culture and the significance of place. *Discourse: Studies in the Cultural Politics of Education* 29(3) 325–42.

Kellett, Mary (2005) *How to develop children as researchers: a step-by-step guide to teaching the research process*. London: Paul Chapman Publishing.

Kemmis, Stephen and McTaggart, Robin (2000) Participatory action research. In Denzin, Norman K. and Lincoln, Yvonna S. (eds) *Handbook of qualitative research* (second edition). London: Sage. 567–606.

Kendrick, Andrew, Steckley, Laura, and Lerpiniere, Jennifer (2008) Ethical issues, research and vulnerability: gaining the views of children and young people in residential care. *Children's Geographies* 6(1) 79–93.

Kenway, Jane (1997) Taking stock of gender reform policies for Australian schools: past, present and future. *British Educational Research Journal* 23(3) 329–44.

Kenway, Jane and Hickey-Moody, Anna (2009) Spatialized leisure-pleasures and masculine distinctions. *Social and Cultural Geography* 10(8).

Kenworthy Teather, Elizabeth (1999) *Embodied geographies: space, bodies and rites of passage*. London: Routledge.

Kenyon, Elizabeth (2003) Young adults' household formation: individualisation, identity and home. In Allan, Graham and Jones, Gill (eds) *Social relations and the life course*. New York: Palgrave Macmillan. 103–19.

Kenyon, Elizabeth and Heath, Sue (2001) Choosing this life: narratives of choice amongst house sharers. *Housing Studies* 16(5) 619–35.

Kimberlee, Richard H. (2002) Why don't British young people vote at general elections. *Journal of Youth Studies* 5(1) 85–98.

Kindon, Sara (2005) Participatory action research. Hay, Iain (ed.) *Qualitative research methods in human geography* (second edition). Oxford University Press: Oxford. 207–20.

Kindon, Sara, Pain, Rachel and Kesby, Mike (2007) Participatory action research: origins, approaches and methods. In Kindon, Sara, Pain, Rachel and Kesby, Mike (eds) *Participatory action research approaches and methods: connecting people, participation and place*. London: Routledge. 9–18.

King, Russell and Ruiz-Gelices, Enric (2003) International student migration and the European 'year abroad': effects on European identity and subsequent migration behaviour. *International Journal of Population Geography* 9 229–52.

Kintrea, Keith, Bannister, Jon, Pickering, Jon, Reid, Maggie and Suzuki, Naofumi (2008) *Young people and territoriality in British cities*. York: Joseph Rowntree Foundation.

Kraack, Anna and Kenway, Jane (2002) Place, time and stigmatised youthful identities: bad boys in paradise. *Journal of Rural Studies* 18(2) 145–55.

Krenichyn, Kira (1999) Messages about adolescent identity: coded and contested spaces in a New York City high school. In Kenworthy Teather, Elizabeth (ed.) *Embodied geographies: spaces, bodies and rites of passage*. London: Routledge. 43–58.

Laegren, Anne Sofie (2002) The petrol station and the Internet café: rural technospaces for youth. *Journal of Rural Studies* 18(2) 157–68.

Lake, Amelia and Townshend, Tim (2006) Obesogenic environments: exploring the built and food environments. *The Journal for the Royal Society for the Promotion of Health* 126 262–7.

Lawler, Steph (2008) *Identity: sociological perspectives*. Cambridge: Polity.

Leonard, Madeleine (2006) Teens and territory in contested spaces: negotiating sectarian influences in Northern Ireland. *Children's Geographies* 4(2) 225–38.

Leyshon, Michael (2002) On being 'in the field': practice, progress and problems in researching young people in rural areas. *Journal of Rural Studies* 18(2) 179–91.

—— (2008a) 'We're stuck in the corner': young women, embodiment and drinking in the countryside. *Drugs: Education, Prevention and Policy* 15(3) 267–89.

—— (2008b) The betweenness of being a rural youth: inclusive and exclusive lifestyles. *Social and Cultural Geography* 9(1) 1–26.

Lincoln, Sian (2005) Feeling the noise: teenagers, bedrooms and music. *Leisure Studies* 24(4) 399–414.

Lindsay, Jo (2009) Young Australians and the staging of intoxication and self-control. *Journal of Youth Studies* 12(4) 371–84.

Longhurst, Robyn (2001) *Bodies: exploring fluid boundaries*. London: Routledge.

—— (2005) Situating bodies. In Nelson, Lisa, and Seager, Joni (eds) *A companion to feminist geography*. Oxford: Blackwell. 337–49.

Mac an Ghaill, Mairtin (1997) *The making of men: masculinities, sexualities and schooling*. Buckingham: Open University Press.

MacDonald, Rob and Marsh, Jane (2004) Missing school. *Youth and Society* 36(2) 143–62.

MacDonald, Robert and Shildrick, Tracy (2007) Street corner society: leisure careers, youth (sub)culture and social exclusion. *Leisure Studies* 26(3) 339–55.

MacDonald, Robert, Shildrick, Tracy, Webster, Colin and Simpson, Donald (2005) Growing up in poor neighbourhoods: the significance of class and place in the extended transitions of 'socially excluded' young adults. *Sociology* 39(2) 873–91.

MacDonald, Robert, Mason, Paul, Shildrick, Tracy, Webster, Colin, Johnston, Les and Ridley, Louise (2001) Snakes and ladders: in defence of studies of youth transition. *Sociological Research Online* 5(4). Available on line at http://www.socresonline.org.uk/5/4/macdonald.html.

Malbon, Ben (1999) *Clubbing: dancing, ecstasy and vitality*. London: Routledge.

Mallett, Shelley (2004) Understanding home: a critical review of the literature. *Sociological Review* 52(1) 62–89.

Mallett, Shelley, Rosenthal, Doreen, Myers, Paul, Milburn, Norweeta, Rotheram-Borus, Mary Jen (2004) Practising homelessness: a typology approach to young people's daily routines. *Journal of Adolescence* 27(3) 337–49.

Malone, Karen (2002) Street life: youth, culture and competing uses of public space. *Local Environment* 14(2) 157–68.

Marsh, David, O'Toole, Therese and Jones, Su (2007) *Young people and politics in the UK: apathy or alienation?* New York: Palgrave Macmillan.

Marston, Sallie (2004) A long way from home: domesticating the social production of scale. Sheppard, Eric and McMaster, Robert B. (eds) *Scale and geographic enquiry: nature, society and method*. Oxford: Blackwell. 170–91.

Mason, Michael, Singelton, Andrew and Webber, Ruth (2007) The spirituality of young Australians. *International Journal of Children's Spirituality* 12(2) 149–63.

Massey, Doreen (1994) *Space, place and gender*. Minneapolis: University of Minnesota Press.

Matthews, Hugh (2001) Power games and moral territories: ethical dilemmas when working with children and young people. *Ethics, Place and Environment* 4(2) 117–78.

—— (2003) The street as a liminal space: the barbed spaces of childhood. In Christensen, Pia and O'Brien, Margaret (eds) *Children in the city: home, neighbourhood and community*. London: Routledge. 101–17.

Matthews, Hugh and Limb, Melanie (1998) The right to say: the development of youth councils/forums within the UK. *Area* 30(1) 66–78.

—— (2003) Another white elephant? Young councils as democratic structures. *Space and Polity* 7(2) 173–92.

Matthews, Hugh, Limb, Melanie and Percy-Smith, Barry (1998) Changing worlds: the microgeographies of young teenagers. *Tijdschrift voor Economische en Sociale Geografie* 89(2) 193–202.

Matthews, Hugh, Limb, Melanie and Taylor, Mark (1999a) Young people's participation and representation in society. *Geoforum* 30 135–44.

—— (1999b) Reclaiming the street: the discourse of curfew. *Environment and Planning A* 31 1713–30.

Matthews, Hugh, Taylor, Mark, Percy-Smith, Barry and Limb, Melanie (2000) The unacceptable flaneur: the shopping mall as a teenage hangout. *Childhood* 7(3) 279–94.

Maxwell, Joseph (1996) *Qualitative research design: an interactive approach*. London: Sage.

McCulloch, Ken, Stewart, Alexis and Lovergreen, Nick (2006) 'We just hang out together': youth cultures and social class. *Journal of Youth Studies* 9(5) 539–56.

McDowell, Linda (2001) 'It's that Linda again': ethical, practical and political issues involved in longitudinal research with young men. *Ethics, Place and Environment* 4(2) 87–100.

—— (2002) Masculine discourses and dissonances: strutting 'lads', protest masculinity, and domestic respectability. *Environment and Planning D: Society and Space* 20 97–119.

—— (2003) Masculine identities and low-paid work: young men in urban labour markets. *International Journal of Urban and Regional Research* 27(4) 828–48.

—— (2005) The men and the boys: bankers, burger makers and barmen. In Van Hoven, Bettina and Hörschelmann, Kathrin (eds) *Spaces of masculinities*. London: Routledge. 19–30.

McGregor, Jane (2004) Editorial. *FORUM* 46(1) 2–5.

McKendrick, John (2001) Coming of age: rethinking the role of children in population studies. *International Journal of Population Geography* 7 461–72.

McNamee, Sara, Valentine, Gill, Skelton, Tracey and Butler, Ruth (2003) Negotiating difference: lesbian and gay transitions to adulthood. In Allan, Graham and Jones, Gill (eds) *Social relations and the life course*. New York: Palgrave Macmillan. 120–34.

Miles, Steven (2000) *Youth lifestyles in a changing world*. Milton Keynes: Open University Press.

Mitchell, Don (2003) *The right to the city: social justice and the fight for public space*. New York: Guildford Press.

Morrow, Virginia (2001) *Network and neighbourhoods: children's and young people's perspectives*. London: Health Development Agency.

—— (2008) Ethical dilemmas in research with children and young people about their social environments. *Children's Geographies* 6(1) 49–61.

Muncie, John (2004) *Youth and crime*. London: Sage.

Nairn, Karen, Higgins, Jane, Thompson, Brigid, Anderson, Megan and Fu, Nedra (2006) 'It's just like the teenage stereotype, you go out and drink and stuff': hearing from young people who don't drink. *Journal of Youth Studies* 9(3) 287–304.

Nayak, Anoop (2003a) *Race, place and globalization: youth cultures in a changing world*. Oxford: Berg.

—— (2003b) White masculinities and the subcultural response to deindustrialisation. *Environment and Planning D: Society and Space* 21 7–25.

—— (2006) Displaced masculinities: chavs, youth and class in the post-industrial city. *Sociology* 40(5) 813–31.

Nespor, Jan (2000) Anonymity and place in qualitative inquiry. *Qualitative Inquiry* 6(4) 546–69.

Noble, Greg (2005) The discomfort of strangers: racism, incivility and ontological security in a relaxed and comfortable nation. *Journal of Intercultural Studies* 26(1) 107–20.

O'Dougherty, Maureen (2006) Public relations, private security: managing youth and race at the Mall of America. *Environment and Planning D: Society and Space* 24 131–54.

Oldman, Joe (2004) Beyond bricks and mortar. In Roche, Jeremy, Tucker, Stanley, Thomson, Rachel and Flynn, Ronny (eds) *Youth in society* (second edition). London: Sage. 112–19.

Olsson, Craig A., Bond, Lydal, Burns, Jane M., Vella-Brodrick, Dianne A. and Sawyer, Susan M. (2003) Adolescent resilience: a concept analysis. *Journal of Adolescence* 26 1–11.

O'Reilly, Camille Caprioglio (2006) From drifter to gap year tourist: mainstreaming backpacker travel. *Annals of Tourism Research* 33(4) 998–1017.

O'Toole, Thérèse (2003) Engaging with young people's conceptions of the political. *Children's Geographies* 1(1) 71–90.

O'Toole, Thérèse, Lister, Michael, Marsh, Dave, Jones, Su and McDonagh, Alex (2003) Tuning out or left out? Participation and non-participation among young people. *Contemporary Politics* 9(1) 45–61.

Owens, Patsy Eubanks (2002) No teens allowed: the exclusion and adolescents from public spaces. *Landscape Journal* 21(1) 156–63.

Pain, Rachel (2000) Place, social relations and the fear of crime: a review. *Progress in Human Geography* 24(3) 365–87.

—— (2001a) Age, generation and lifecourse. In Barke, Michael, Fuller, Duncan, Gough, Jamie, MacFarlane, Robert and Mowl, Graham (2001) *Introducing Social Geographies*. London: Arnold. 141–63.

—— (2001b) Gender, race, age and fear in the city. *Urban Studies* 5–6 899–913.

—— (2003) Youth, age and the representation of fear. *Capital and Class* 60 151–71.

—— (2006) Paranoid parenting? Rematerializing risk and fear for children. *Social and Cultural Geography* 7(2) 221–43.

—— (2009) Globalized fear? Towards an emotional geopolitics. *Progress in Human Geography* 33(4) 466–86.

Pain, Rachel and Francis, Peter (2004) Living with crime: spaces of risk for homeless young people. *Children's Geographies* 2(1) 95–110.

Pain, Rachel and Hopkins, Peter (2009) Social geographies of age and ageism: landscapes, lifecourses and justice. In Smith, Susan J., Pain, Rachel, Marston, Sallie, and Jones III, John Paul (eds) *Handbook of social geography*. London: Sage.

Pain, Rachel and Smith, Susan J. (eds) (2008) *Fear: critical geopolitics and everyday life*. Aldershot: Ashgate.

Pain, Rachel, Gough, Jamie, Mowl, Graham, Barke, Michael, MacFarlane, Robert and Fuller, Duncan (2001) *Introducing social geographies*. London: Arnold.

Pain, Rachel, Grundy, Sue and Gill, Sally with Towner, Elizabeth, Sparks, Geoff and Hughes, Kate (2005) 'So long as I take my mobile': mobile phones, urban life and geographies of young people's safety. *International Journal of Urban and Regional Research* 29(4) 824–30.

Panelli, Ruth (2002) Young rural lives: strategies beyond diversity. *Journal of Rural Studies* 18(2) 113–22.

—— (2004) *Social geographies*. London: Sage.

Park, Alison (2004) Has modern politics disenchanted the young? In Park, Alison, Curtice, John, Thomson, Katarina, Bromley, Catherine and Phillips, Miranda (eds) *British social attitudes*. London: Sage.

Peek, Lori (2003a) Reactions and response: Muslim students' experiences on New York City campuses post 9/11. *Journal of Muslim Minority Affairs* 23(2) 271–83.

—— (2003b) Community isolation and group solidarity: examining the Muslim student experience after September 11 2001. In Monday, Jacquelyn L. (ed.) *Beyond September 11: an account of post-disaster research*. Boulder: University of Colorado. 333–54.

Penrose, Jan (1995) Essential constructions? The 'cultural bases' of nationalist movements. *Nations and Nationalism* 1(3) 391–417.

Percy-Smith, Barry and Matthews, Hugh (2001) Tyrannical spaces: young people, bullying and urban neighbourhoods. *Local Environment* 6(1) 49–63.

Peterson, Abby (2000) 'A locked door' – the meaning of home for punitively homeless young people in Sweden. *Young* 8(2) 24–44.

Philo, Chris (2001) Accumulating populations: bodies, institutions and space. *International Journal of Population Geography* 7 473–90.

—— (2003) 'To go back up the side hill': memories, imaginations and reveries of childhood. *Children's Geographies* 1(1) 7–23.

Philo, Chris and Parr, Hester (2000) Institutional geographies: introductory remarks. *Geoforum* 31 513–21.

Philo, Chris and Smith, Fiona M. (2003) Guest editorial: political geographies of children and young people. *Space and Polity* 7(2) 99–115.

Phoenix, Ann and Pattynama, Pamela (2006) Intersectionality. *European Journal of Women's Studies* 13(3) 187–92.

Plows, Vicky and Gallagher, Michael (2009) Evaluation of a youth counselling service: surveying young people's views. In Tisdall, Kay, David, John and Gallagher, Michael (eds) *Researching with children and young people: research design, methods and analysis*. London: Sage. 97–104.

Punch, Samantha (2002) Research with children: the same or different from research with adults? *Childhood* 9(3) 321–41.

Quintelier, Ellen (2008) Who is politically active: the athlete, the scout member or the environmental activist? Young people, voluntary engagement and political participation. *Acta Sociologica* 51(4) 355–70.

Rawlins, Emma (2008) Citizenship, health education and the 'obesity epidemic'. *ACME* 7(2) 135–51.

Redshaw, Sarah and Noble, Greg (2006) *Mobility, gender and young drivers: second report of the transforming drivers study of young people and driving*. Sydney: National Roads and Motorist's Association.

Richard, Birgit (2006) Goth. In Steinberg, Shirley R., Parmar, Priya and Richard, Birgit (eds) *Contemporary youth culture: an international encyclopaedia*. Westport, CT: Greenwood Press. 431–8.

Ritzer, George (1993) *The McDonaldization of society*. Thousand Oaks, CA: Pine Forge Press.

Robb, Martin (ed.) (2007) *Youth in context: frameworks, settings and encounters*. London: Sage.

Roberts, Ken (2007) Youth transitions and generations: a response to Wyn and Woodman. *Journal of Youth Studies* 10(2) 263–9.

Rosenthal, Doreen and Rotheram-Borus, Mary Jane (2005) Editorial: young people and homelessness. *Journal of Adolescence* 28(2) 167–9.

Rowlands, Rob and Gurney, Craig M. (2000) Young people's perceptions of housing tenure: a case study in the socialization of tenure prejudice. *Housing, Theory and Society* 17(3) 121–30.

Ruddick, Sue (1996) *Young and homeless in Hollywood: mapping social identities*. London: Routledge.

—— (2003) The politics of aging: globalisation and the restructuring of youth and childhood. *Antipode* 35(2) 334–62.

Rugg, Julie (1999) *Young people, housing and social policy*. London: Routledge.

Runnymede Trust (1998) *Young people in the UK: attitudes and opinions on Europe, Europeans and the European Union*. Runnymede Trust in partnership with the Commission for Racial Equality.

Save The Children (2004) *Separated children in Europe programme: statement of good practice* (third edition). London: SCF.

Saville, Stephen John (2008) Playing with fear: parkour and the mobility of emotion. *Social and Cultural Geography* 9(8) 891–914.

Scanlon, Lesley, Rowling, Louise and Weber, Zita (2007) 'You don't have like an identity . . . you're just lost in a crowd': forming a student identity in the first-year transition to university. *Journal of Youth Studies* 10(2) 223–41.

Scott, Alister, Gilbert, Alana and Gelan, Ayele (2007) *The urba–rural divide: myth or reality*. Aberdeen: Macaulay Institute.

Scourfield, Jonathan, Davies, Andrew and Holland, Sally (2004) Wales and Welshness in middle childhood. *Contemporary Wales* 16 83–100.

Scourfield, Jonathan, Dicks, Bella, Drakeford, Mark and Davies, Andrew (2006a) *Children, place and identity: nation and locality in middle childhood*. London: Routledge.

Scourfield, Jonathan, Dicks, Bella, Holland, Sally, Drakeford, Mark and Davies, Andrew

(2006b) The significance of place in middle childhood: qualitative evidence from Wales. *British Journal of Sociology* 57(4) 577–95.

Shaw, Ian and Warf, Barney (2009) Worlds of affect: virtual geographies of video games. *Environment and Planning A* 41(6) 1332–43.

Sheller, Mimi and Urry, John (2006) The new mobilities paradigm. *Environment and Planning A* 38 207–26.

Shildrick, Tracy (2006) Youth culture, subculture and the importance of neighbourhood. *Young* 14(1) 61–74.

Shildrick, Tracy and MacDonald, Robert (2005) In defence of subculture: young people, leisure and divisions. *Journal of Youth Studies* 9(2) 125–40.

Short, John Rennie and Kim, Yeong-Hyun (1999) *Globalisation and the city*. Michigan: Longman.

Shucksmith, Mark (2004) Young people and social exclusion in rural areas. *Sociologia Ruralis* 44(1) 43–59.

Sibley, David (1995) Families and domestic routines: constructing the boundaries of childhood. Pile, Steve and Thrift, Nigel (eds) *Mapping the subject: geographies of cultural transformation*. London: Routledge. 123–37.

Simpson, Kate (2004) 'Doing development': the gap year, volunteer-tourists and a popular practice of development. *Journal of International Development* 16 681–92.

—— (2005) Dropping out or signing up? The professionalisation of youth travel. *Antipode* 37(3) 447–69.

Skeggs, Beverley (1997) *Formations of class and gender*. London: Sage.

Skelton, Tracey (2000) 'Nothing to do, nowhere to go?' Teenage girls and 'public' solace in the Rhondda Valleys, South Wales. In Holloway, Sarah and Valentine, Gill (eds) *Children's geographies playing, living, learning*. London: Routledge. 80–99.

—— (2001) Girls in the club: researching working class girls' lives, *Ethics, Place and Environment* 4(2) 167–73.

—— (2008) Research with children and young people: exploring the tensions between ethics, competence and participation. *Children's Geographies* 6(1) 21–36.

—— (2009) Children's geographies/geographies of children: play, work, mobilities and migration. *Geography Compass* 3(4) 1430–48.

Skelton, Tracey and Valentine, Gill (eds) (1998) *Cool places: geographies of youth cultures*. London: Routledge.

—— (2003) Political participation, political action and political identities: young deaf people's perspectives. *Space and Polity* 7(2) 117–34.

—— (2005) Exploring notions of masculinity and fatherhood: when gay sons 'come out' to heterosexual fathers. In Van Hoven, Bettina and Hörschelmann, Kathrin (eds) *Spaces of masculinities*. London: Routledge. 207–21.

Smith, Christian (2003) Theorizing religious effects among American adolescents. *Journal for the Scientific Study of Religion* 42(1) 17–30.

Smith, Christian, Denton, Melinda Lundquist, Faris, Robert and Regnerus, Mark (2002) Mapping American adolescent religious participation. *Journal for the Social Scientific Study of Religion* 14(4) 597–612.

Smith, Darren (2005) 'Studentification: the gentrification factory? In Atkinson, Rowland and Bridge, Gary (eds) *Gentrification in the global context: the new urban colonialism*. London: Routledge. 72–89.

—— (2009) 'Student geographies', urban restructuring, and the expansion of higher education. *Environment and Planning A* 41 1795–1804.

Smith, Darren and Holt, Louise (2007) Studentification and 'apprentice' gentrifiers within Britain's provincial towns and cities: extending the meaning of gentrification. *Environment and Planning A* 39 142–61.

Smith, Neil (1993) Homeless/global: scaling places. In Bird, Jon, Curtis, Brian, Putnam, Tim, Robertson, George and Tickner, Lisa (eds) *Local cultures, global change*. London: Routledge.

Smith, Susan, J. (1993) Social landscapes: continuity and change. In Johnston, Ron (ed.) *A Changing world: a changing discipline*. Oxford, Blackwell.

—— (1999) Society-space. In Cloke, Paul, Crang, Philip and Goodwin, Mark (eds) *Introducing human geographies*. London: Arnold. 212–23.

—— (2005) Joined-up geographies. *Transactions of the Institute of British Geographers* 30 329–90.

Somerville, Peter (1997) The social construction of home. *Journal of Architectural and Planning Research* 14(3) 26–245.

Stratford, Elaine (2002) On the edge: a tale of skaters and urban governance. *Social and Cultural Geography* 3(2) 193–206.

Such, Elizabeth, Walker, Oliver and Walker, Robert (2005) Anti-war children: representation of youth protests against the Second Iraq War in the British national press. *Childhood* 12(3) 301–25.

Thomas, Mary E. (2000) From crib to campus: kids' sexual/gender identities and institutional space. *Environment and Planning D: Society and Space* 32 577–80.

—— (2005) Girls, consumption space and the contradictions of hanging out in the city. *Social and Cultural Geography* 6(4) 587–605.

Thompson, Andrew (2001) Nations, national identities and human agency: putting people back into nations. *Sociological Review* 49(2) 18–32.

Thompson, Lee and Cupples, Julie (2008) Seen and not heard? Text messages and digital sociality. *Social and Cultural Geography* 9(1) 95–108.

Thomson, Rachel (2007) A biographical perspective. Kehily, Mary Jane (ed.) *Understanding youth: perspectives, identities and practices*. Sage: London. 73–106.

Tisdall, Kay, Davis, John and Gallagher, Michael (2009) *Researching with children and young people: research design, methods and analysis*. London: Sage.

Tucker, Faith (2003) Sameness or difference? Exploring girls' use of recreational spaces. *Children's Geographies* 1(1) 111–24.

Tucker, Faith and Matthews, Hugh (2001) 'They don't like girls hanging around there': conflicts over recreational space in rural Northamptonshire. *Area* 33(2) 161–8.

Turner, Sarah and Manderson, Desmond (2007) Socialisation in a space of law: student performativity at 'Coffee House' in a university law faculty. *Environment and Planning D: Society and Space* 25 761–82.

Tutenges, Sebastian and Hulvej Rod, Morten (2009) 'We got incredibly drunk . . . it was dammed fun': drinking stories among Danish youth. *Journal of Youth Studies* 12(4) 355–170.

Tyler, Imogen (2008) 'Chav mum, chav scum': class disgust in contemporary Britain. *Feminist Media Studies* 8(1) 17–34.

UNHCR (2004) *Trends in unaccompanied and separated children seeking asylum in industrialized countries, 2001–2003*. Geneva: UNHCR.

—— (2008a) *Global refugee trends: overview of refugee populations, new arrivals, durable solutions, asylum-seekers, stateless and other persons of concern to the UNHCR, 2007*. Geneva: UNHCR.

—— (2008b) *Asylum levels and trends in industrialized countries: first half of 2008*. Geneva: UNHCR.

Vaaranen, Heli and Weiloch, Neil (2002) Car crashes and dead end careers: leisure pursuits of the Finnish subculture of the Kortteliralli Street racing. *Young* 10(1) 42–58.

Valentine, Gill (1996) Angels and devils: moral landscapes of childhood. *Environment and Planning D: Society and Space* 14 581–99.

—— (1997) 'Oh yes I can.' 'Oh no you can't': children's and parents understandings of kids' competence to negotiate public space safely. *Antipode* 29(1) 65–89.

—— (1999) Being seen and heard? The ethical complexities of working with children and young people at home and at school. *Ethics, Place and Environment* 2(2) 141–55.

—— (2001) *Social geographies*. Harlow: Pearson Education.

—— (2003) Boundary crossings: transitions from childhood to adulthood. *Children's Geographies* 1(1) 37–52.

Valentine, Gill, Butler, Ruth and Skelton, Tracey (2001) The ethical and methodological complexities of doing research with 'vulnerable' young people. *Ethics, Place and Environment* 4(2) 119–25.

Valentine, Gill, Holloway, Sarah, Knell, Charlotte and Jayne, Mark (2008) Drinking places: young people and cultures of consumption in rural environments. *Journal of Rural Studies* 24(1) 28–40.

Valentine, Gill, Skelton, Tracey and Chambers, Deborah (1998) Cool places: an introduction to youth and youth cultures. In Skelton, Tracey and Valentine, Gill (eds) *Cool places: geographies of youth cultures*. London: Routledge. 1–32.

Vanderbeck, Robert (2007) Intergenerational geographies: age relations, segregation and reengagements. *Geography Compass* 1 200–21.

—— (2008) Reaching critical mass? Theory, politics and the culture of debate in children's geographies. *Area* 40(3) 393–400.

Vanderbeck, Robert and Dunkley, Cheryl Morse (2003) Young people's narratives of rural–urban difference. *Children's Geographies* 1(2) 241–59.

Vanderbeck, Robert and Johnson, James (2000) 'That's the only place where you can hang out': urban young people and the space of the mall. *Urban Geography* 21(1) 5–25.

Verma, Rita (2006) Trauma, cultural survival and identity politics in a post 9/11 era: reflections by Sikh youth. *Sikh Formations* 2(1) 89–101.

Walsh, Charlotte (2002) Curfews: no more hanging around. *Youth Justice* 2(2) 70–81.

—— (2008) The Mosquito: a repellent response. *Youth Justice* 8(2) 122–33.

Wasoff, Fran, Jamieson, Lynn and Smith, Adam (2005) Solo living, individual and family boundaries: findings from secondary analysis. In McKie, Linda, Cunningham-Burley, Sarah and Campling, Jo (eds) *Families in society: boundaries and relationship*. Bristol: Policy Press. 207–44.

Waters, Johanna, L. (2006a) Geographies of cultural capital: education, international migration and family strategies between Hong Kong and Canada. *Transactions of the Institute of British Geographers* 31 179–92.

—— (2006b) Emergent geographies of international education and social exclusion. *Antipode* 38(5) 1046–8.

Watt, P. (1998) Going out of town: youth, race and place in the south east of England. *Environment and Planning D: Society and Space* 16(6) 687–703.

White, Clarissa, Bruse, Sara and Ritchie, Jane (2000) *Young people's politics: political interest and engagement amongst 14–24 year olds*. York: Joseph Rowntree Foundation.

White, Rob and Wyn, Johanna (2008) *Youth and society*. Oxford: Oxford University Press.

Widdicombe, Sue and Wooffitt, Robin (1995) *The language of youth subcultures: social identity in action*. Hemel Hempstead: Harvester Wheatsheaf.

Willis, Paul (1977) *Learning to labour: how working class kids get working class jobs*. Farnborough: Saxon House.

Wills, Jane (2005) Globalization and protest. In Cloke, Paul, Crang, Philip and Goodwin, Mark (eds) *Introducing human geographies*. London: Arnold. 573–87.

Winchester, Hilary P.M. (2005) Qualitative research and its place in human geography. Hay, Iain (ed.) *Qualitative research methods in human geography* (second edition). Oxford University Press: Oxford. 3–18.

Winchester, Hilary P.M., McGuirk, Pauline M. and Everett, Kathryn (1999) Schoolies week as a rite of passage: a study of celebration and control. In Kenworthy Teather, Elizabeth (ed.) *Embodied Geographies: spaces, bodies and rites of passage*. London: Routledge. 59–77.

Woodhead, Linda and Heelas, Paul (2005) *The spiritual revolution: why religion is giving way to spirituality*. Oxford: Blackwell.

Woolley, Helen and Johns, Ralph (2001) Skateboarding: the city as a playground. *Journal of Urban Design* 6(2) 211–30.

Wyn, Johanna and White, Rob (1997) *Rethinking youth*. Australia: Allen & Unwin.

Wyn, Johanna and, Woodman, Dan (2006) Generation, youth and social change in Australia. *Journal of Youth Studies* 9(5) 495–514.

—— (2007) Researching youth in a time of change: a reply to Roberts. *Journal of Youth Studies* 10(3) 373–81.

Wyness, Michael (2006) *Childhood and society: an introduction to the sociology of childhood*. New York: Palgrave Macmillan.

Zaal, Mayida, Salah, Tahani and Fine, Michelle (2007) The weight of the hyphen: freedom, fusion and responsibility embodied by young Muslim-American women during a time of surveillance. *Applied Development Science* 11(3) 164–77.

Index